Introduction to the Non-Destructive Testing of Welded Joints

Introduction to the Non-Destructive Testing of Welded Joints

R HALMSHAW

A BINGTON PUBLISHING

Woodhead Publishing Ltd in association with The Welding Institute
Cambridge England

Published by Abington Publishing,
Woodhead Publishing Ltd, Abington Hall, Abington,
Cambridge CB1 6AH, England

First published 1988, The Welding Institute
Reprinted 1992, Abington Publishing

ISBN   1 85573 091 X

Printed by Crampton & Sons Limited, Sawston, Cambridge CB2 4BQ, UK

Contents

Foreword

This book has been prepared to give inspecting engineers, welding engineers, designers, producers and purchasers of welded plant, an up-to-date understanding of non-destructive testing (NDT) methods and their scope of application to welds and welded structures.

Non-destructive testing is a technology which is continually advancing and it is necessary, at intervals, to update any published information on the subject. Even the well established techniques such as magnetic crack detection or radiography show considerable advances when considered over, say, ten-year intervals, and in the same period completely new techniques can pass from the research laboratory phase into commercial industrial use.

Because fabrication techniques such as welding require considerable skill, it is inevitable that mistakes will occur, that flaws will be incorporated into the material, and that therefore some post-fabrication testing will be essential to discover any such serious flaws which may impair performance. Today, also, there is a new emphasis on in-service inspection of fabricated structures and on structure monitoring. Non-destructive testing is an inspection process which forms part of the whole quality assurance/quality control (QA/QC) scheme.

Most people working in NDT in the United Kingdom have trained as engineers, metallurgists or physicists, and have learnt their NDT by experience rather than by formal training. Very few people have had a formal training in a wide range of NDT methods, and most, therefore, tend to be knowledgeable only about the techniques used in their own industries. This book describes all methods of NDT relevant to weld inspection, and gives a picture of what is involved — the difficulties, the advantages and the limitations — in as unbiased a manner as possible.

Some NDT methods are relatively simple to apply whereas others require considerable skill, and while this book does not intend to be a code of good practice, it should provide sufficient detail for the reader to form a clear understanding of the requirements of each technique for successful application. Welded constructions vary so much in type, design and size that only very general suggestions can be made about the suitability of application of any particular technique.

There are many very strongly held opinions on NDT. For example, there are as many sceptics of the usefulness of acoustic emission (AE) methods as there are those who think that AE is the most important development in NDT for years. Many supporters of ultrasonic testing think that radiography cannot find cracks and is an outdated technique, whereas those who advocate radiography believe that ultrasonic testing is unreliable unless done by extremely skilled operators. It is hoped that this book will appeal to practitioners of NDT who need to know what alternative methods exist, and want an unbiased opinion of their advantages and disadvantages. With some of the newer methods of NDT which are included an attempt has been made to indicate the likely direction of future developments.

R. Halmshaw

1   Visual methods

Visual methods of examination, with or without optical aids, are not used as fully as possible by NDT personnel. Gross surface defects, such as severe undercut or incompletely filled grooves, can lead to immediate rejection before any more expensive testing is undertaken. The general appearance of a weld surface can sometimes provide information about the weld quality and a metallographic examination of the structure of the weld metal and the surrounding area is sometimes made by means of a replica technique.[1] Defects such as misalignment, weld globules, shrinkage grooves and incorrect grinding are easily seen.

Visual inspection also includes the measurement of various weld parameters such as surface contour, and visual inspections should be made before welding to check on the form of preparation, cleanliness, fit-up, etc.[2]

Optical aids to visual inspection

Aids to visual inspection should be used whenever practicable (Fig.1). For local examination of a portion of weld that is directly visible to the eye, a small hand lens (magnification 2-4×) used in conjunction with a pen torch is very useful.[3]

The 'borescope' or 'intrascope', an optical system with its own light source, allows the operator to see a magnified view of otherwise inaccessible surfaces and can be used to examine welds on the internal surfaces of tubes or inserted into holes in machinery. The design of the borescope has been transformed in recent years by the application of fibre-optical systems, which can illuminate a surface and retrieve an image over distances up to several metres.

In all remote viewing systems the direction of light incident on the surface being examined is crucially important. If possible, this should be capable of being varied so that detail can be shown in slight relief, and glare and dazzle effects avoided. One problem often encountered is that it is difficult to identify the precise area of the subject which is in view, and careful precautions are required to overcome this.

Desirable features in a borescope are that it should have as large a field of vision as possible, minimum image distortion and adequate illumination. Borescopes are now available which are coupled to a closed-circuit television (CCTV) camera, and sub-miniature TV cameras have been built for insertion into small tubes.

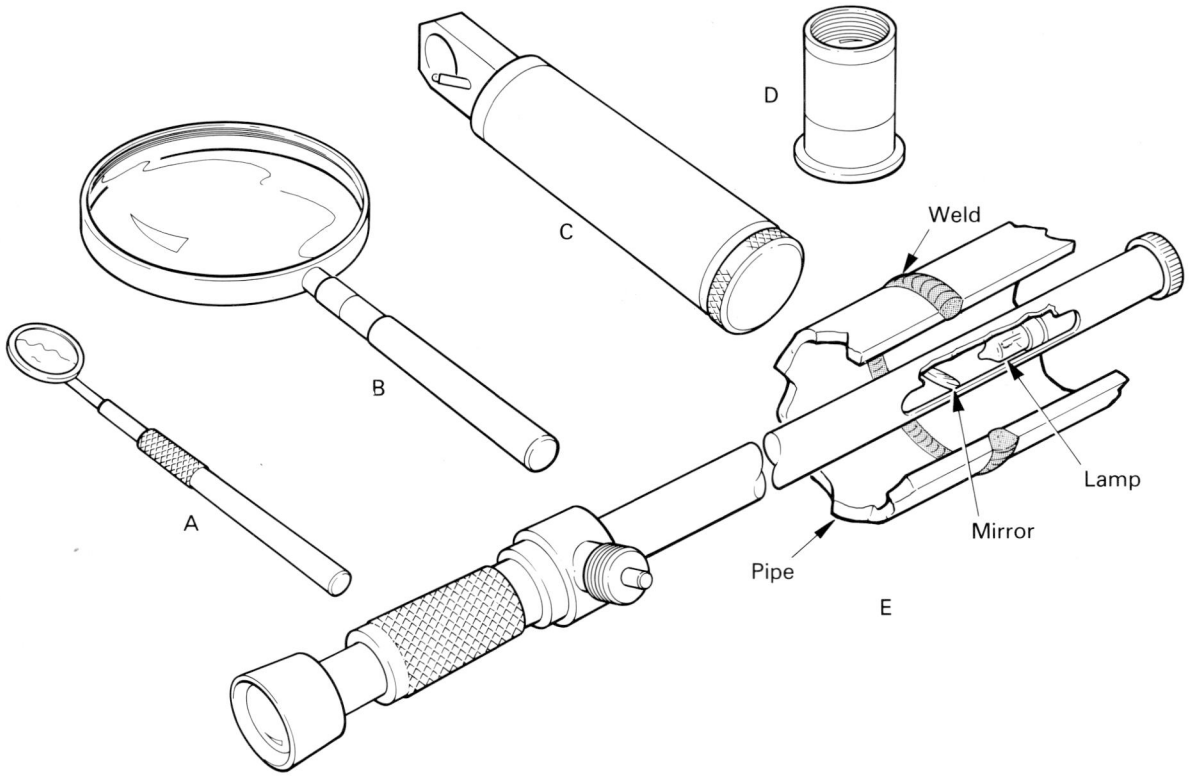

1 Optical aids to visual inspection:
A mirror on stem: may be flat for normal view or concave for limited magnification;
B hand magnifying glass (magnification usually 2-3×);
C illuminated magnifier: field of view more restricted than D (magnification 5-10×);
D inspection glass, usually fitted with a scale for measurement: the front surface is placed in contact with the work (magnification 5-10×);
E borescope or intrascope with built-in illumination (magnification 2-3×).

Additional accessories

Apart from the obvious requirements such as a weld size gauge, ruler, protractor and calipers, straight-edge special devices such as contour gauges and templates are necessary. Modelling clay, Plasticine and cold-setting rubber can be applied to determine the weld profile in regions where access for a measuring device is restricted. The replica is then measured after removal.

To measure the key dimensions of weld preparation and completed welds, The Welding Institute supplies a weld gauge which measures angle of preparation, fillet weld throat size, fillet weld leg length, and height of excess weld metal. The gauges can also be used to measure undercut and misalignment of parts (Fig.2).

Visual examination after welding

The following details should be checked after welding is completed:

· **Cleaning and dressing:** check that all weld slag has been removed. When the weld is specified to be dressed, ensure that the weld metal merges smoothly into the parent metal without under-flushing.

2 Using The Welding
Institute weld gauge to
measure:

A angle of preparation;
B fillet weld throat size;
C fillet weld leg length;
D height of excess weld
    metal.

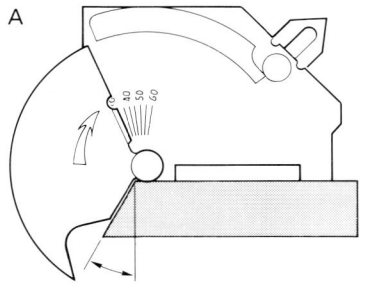

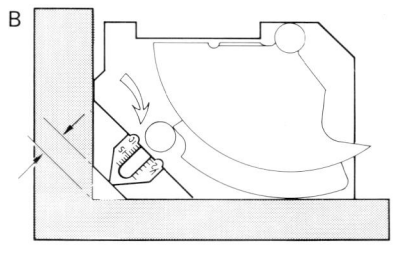

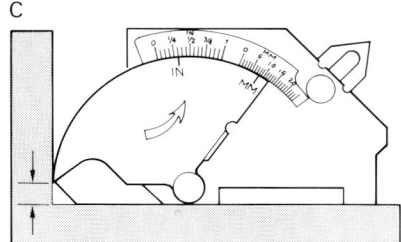

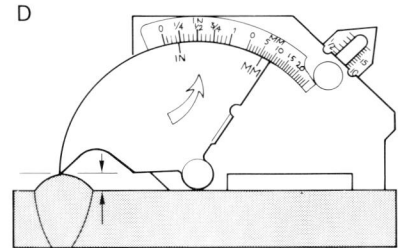

· **Penetration and root examination:** check that over the whole of the joint the penetration and any root concavity, burn-through, or shrinkage grooves are within acceptable limits.
· **Contour:** check that the contour of the weld face and height of the excess weld metal meet the acceptance criteria. Check that the surface of the weld is regular and that weld weave marks are even.
· **Weld width:** check that this is consistent over the whole length of the weld. In fillet welds it is more probable that the throat thickness has been specified; this can be calculated from the width or the profile.
· **Undercut:** measure this with a weld gauge, or where a high degree of accuracy is required, use a surface contour gauge or a depth gauge and compare with acceptance criteria.
· **Overlap:** if visible, this is usually unacceptable.
· **Stray arcing:** this can cause cracks and local hard spots.
· **Weld flaws:** cracks can often be seen visually or with surface crack detection techniques before more expensive NDT such as ultrasonic testing or radiographic methods are applied. (Surface crack detection techniques are described in Chapter 5.)

References

1   BS 5166: 1974. Method for metallographic replica techniques of surface examination (= ISO 3057: 1974). Publ British Standards Institution, London, 1974.
2   BS 5289: 1983. Code of Practice. Visual inspection of fusion welded joints. Publ British Standards Institution, London, 1983.
3   BS 5165: 1974. Guide to the selection of low-power magnifiers used in visual inspection. Publ British Standards Institution, London, 1974.

2   Radiographic methods

General principles

X-rays and gamma-rays are electromagnetic radiation of the same physical nature as visible light, infrared, ultraviolet (UV) and radio waves, but they have a wavelength that enables them, to some extent, to penetrate all materials. They are progressively absorbed as they pass through the material and they travel in straight lines. They affect a photographic plate so that, if a source of X-rays is placed on one side of a specimen and a sheet of photographic film on the other (Fig.3), because more radiation will pass through a region of the specimen where there is a cavity than through solid material, this difference in intensity will be

3 Typical set-up for taking radiographs of a weld in flat plate showing: source of radiation, cavity, film, intensifying screens and anti-back-scatter backing.

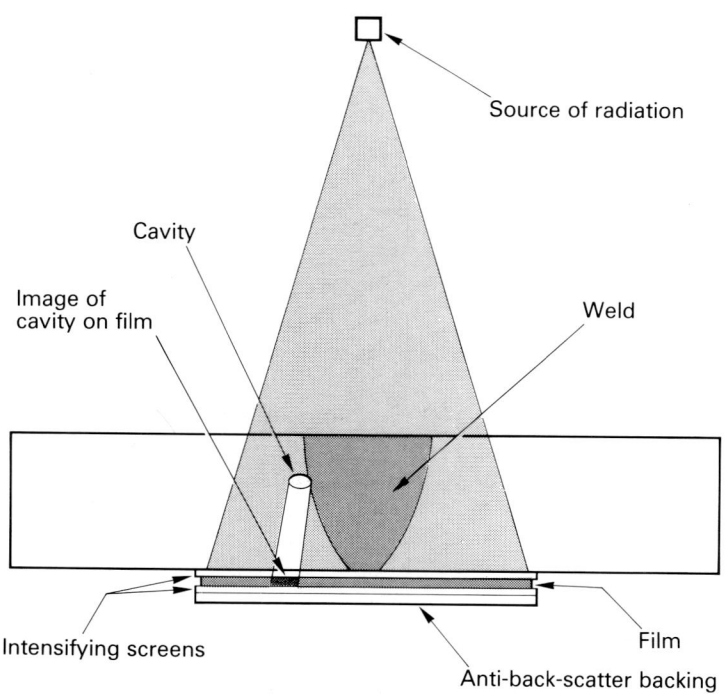

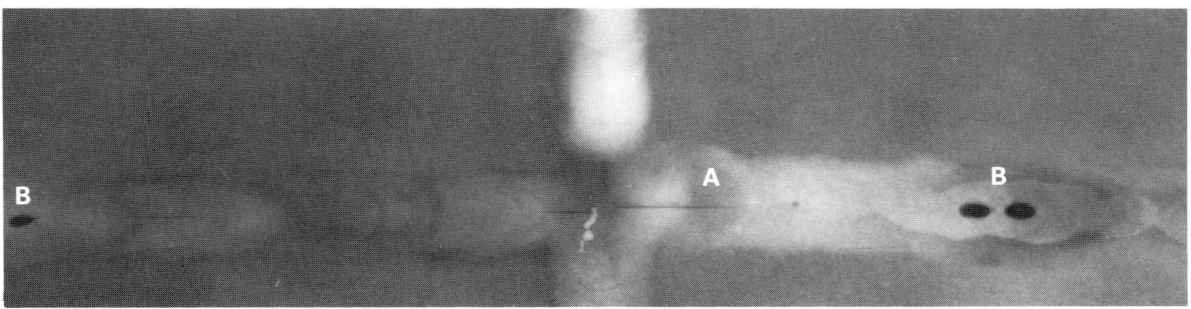

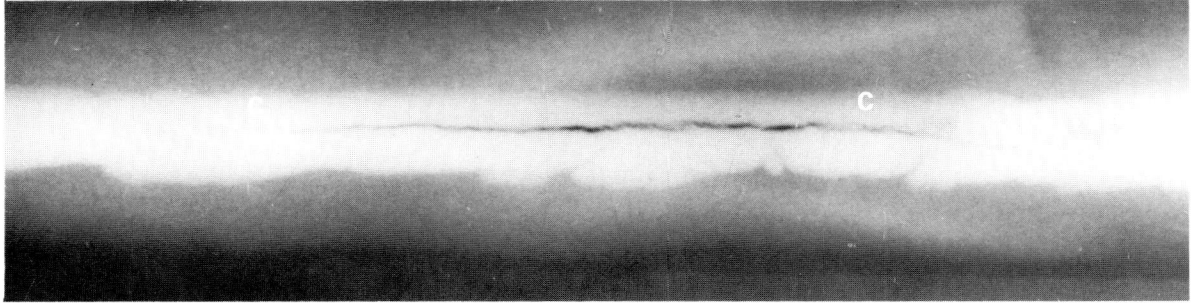

4 Radiographs of welds showing:
A lack of root penetration;
B deep gas cavities;
C longitudinal crack.

recorded on the photographic film; the film, after processing, will show a darker image area under the cavity (Fig.4); and the image will be roughly the same size as the cross-section of the cavity.

As the source of X-rays is normally a near-point source and is spaced away from the specimen, the image is essentially a 'shadow picture' of the internal structure, and the image of the cavity will be a projection of the three-dimensional cavity on to the two-dimensional plane of the film. If the cavity extends a considerable distance through the specimen, the differences in the amount of X-radiation penetrating this region, compared with a solid region of the specimen, will be greater, and the image on the film (after photographic processing) will be correspondingly blacker compared with that of a shallower cavity of the same area. Generally, a simple radiograph will not provide information on the through-thickness location of a cavity (although stereoradiography is possible). An exactly similar argument can be stated for an inclusion in a specimen instead of a cavity, provided this inclusion has a different X-ray absorption from that of the main material of the specimen.

To produce a radiograph, therefore, requires, first of all, an appropriate source of X- or gamma-rays, with the means of switching this on and off for a predetermined exposure time; and, secondly, a photographic film (special films with thick emulsions are used in radiography as these give better results in shorter times). The film needs to be the same size as the specimen and must be held in a light-tight holder with a thin front (the cassette or film envelope) to protect it from light but allow it to be exposed to the X-rays.

Means of processing the film are also required: the conventional photographic technique of develop, fix, wash and dry is usually used. Automatic film processors are available in which the film is inserted at one end, transported through various solutions by rollers, and comes out at the other end processed and dried. Finally, to 'read' the film, it must be placed on an illuminated screen of appropriate luminance (brightness). The quality of this viewing screen light is vitally important and will be described in detail later.

Gamma-rays are electromagnetic radiation, exactly the same as X-rays, but have a different origin. All the above general principles of producing a radiograph apply equally to gamma-radiography.

Neutrons, electrons and protons are also used to produce photographic images analogous to radiographs for special applications, and terms such as 'neutron radiography' are in common use even though neutrons, etc. are atomic particles and not electromagnetic radiation.

Physical principles

X-rays

X-rays are generated when high velocity electrons are stopped by hitting a metal target, and the usual X-ray source is the X-ray tube, consisting of a glass envelope or metal-ceramic vessel having a filament source of electrons at one end (the cathode) and a heavy metal target (the anode) at the other (Fig.5). The X-ray tube is evacuated to a very low gas pressure and may be either sealed or operated on a vacuum pump, most conventional tubes being sealed. A small voltage is applied across the filament to generate electrons and a much larger voltage is applied between the cathode and anode. The X-rays generated are heterogeneous, with a continuous spectrum of complex shape, and various methods are used to describe them.

Roughly, the X-ray region of the electromagnetic spectrum extends over wavelengths from 10^{-7} to 10^{-10}cm, but in industrial radiography wavelength is rarely used to characterise X-rays. It is more general to describe them by the voltage which is applied across the X-ray tube: for example, 100kV X-rays (1kV being 1000 volts). The shortest wavelength in the spectrum of 100kV X-rays will be about 1.2×10^{-9}cm, but the precise shape of the spectrum and the relative intensities of the different wavelengths in the spectrum are not necessarily the same for all 100kV X-ray sets. This is not, therefore, a physically accurate method of describing the X-rays, as the spectrum emitted will depend on whether

5 Two types of X-ray tubes:
A glass envelope;
B metal-ceramic.

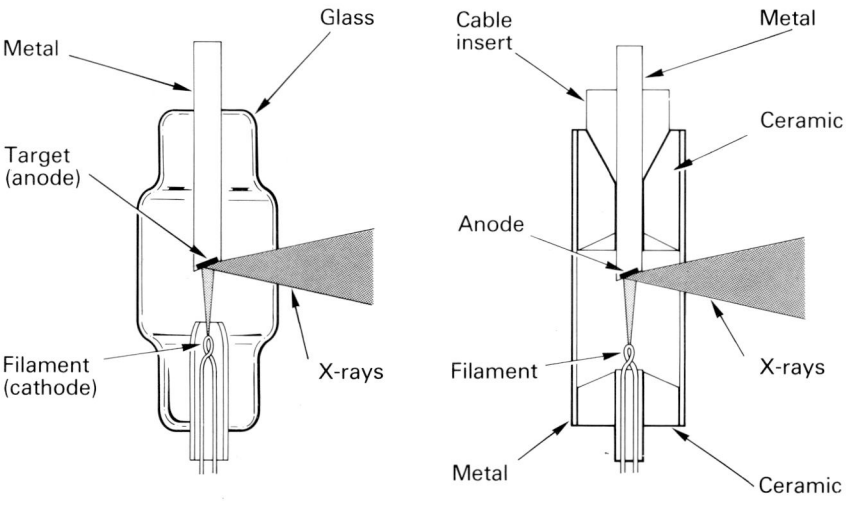

A

B

the applied voltage is constant or varying. Also, the thickness of the wall of the X-ray tube (the window) will affect the output spectrum. Imprecise terms such as: 'soft' X-rays (20—60kV), 'hard' X-rays (150—400kV), and 'very hard' X-rays (400kV upwards) are also used to describe the quality of the radiation.

Industrial radiography uses X-rays from about 20kV to 30MV, according to the thickness of the weldment which has to be penetrated and its material. The intensity of output of an X-ray tube is occasionally given in absolute units of X-ray intensity — for example, roentgens per minute at 1 metre (R/min at 1m), but more usually in terms of the current across the X-ray tube, so that the maximum output of a typical X-ray tube would be given as, for example, 100kV 20mA or 400kV 10mA.

The third basic property of an X-ray tube is the effective width of the source of X-rays — usually called the X-ray focus or focal spot — as seen from a point on the film (Fig.6). By having a sloping anode face the electron beam impinges on a much larger area of the anode, which simplifies the problem of dissipating the heat generated on the anode and allows much larger tube currents to be used without overheating. Conventional X-ray tubes have *effective* focal spot sizes in the range 2×2mm to 5×5mm; and some tubes have a dual focal spot size and include a fine focus facility, say, 0.5×0.5mm. Certain special tubes are made in which the electron beam is focused to much smaller sizes (10-50μm); these are generally known as microfocus tubes and have a much lower output of X-rays because of heat dissipation problems. The importance of focal spot size will be discussed later. (In spite of the use of the term 'X-ray focus', it is important to bear in mind that X-rays cannot be focused: it is the electrons which are focused on to the anode.)

Gamma-rays

Gamma-rays are produced during the decay of some radioactive materials. Radioactive substances may be naturally occurring (radium, uranium, etc.) or artificial radioisotopes, of which there are a very large number. As these decay they may emit alpha-, beta- or gamma-rays or neutrons, or a combination of these. Artificial radioisotopes are either made in atomic reactors by irradiation of suitable material or extracted from spent reactor fuel elements.

6 The effective width of the X-ray tube focal spot (focus) is always measured as seen from a point on the film.

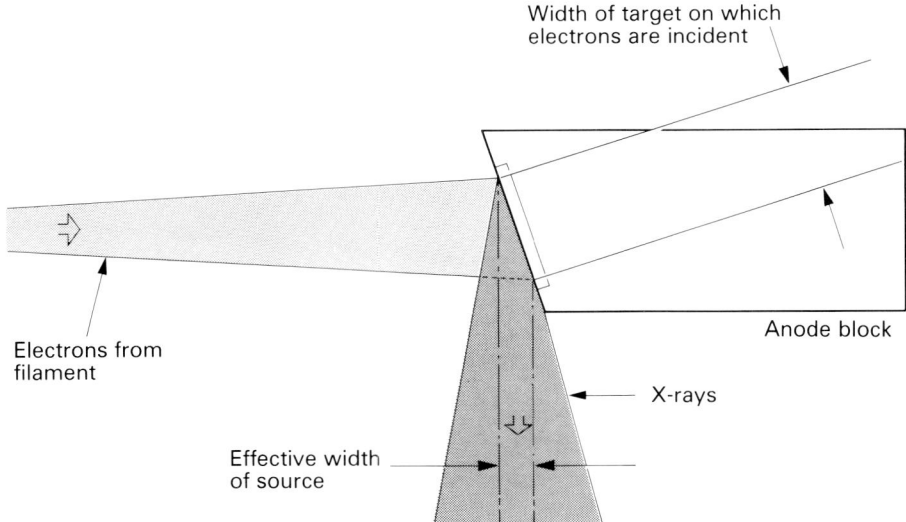

Width of target on which electrons are incident

Electrons from filament

Anode block

X-rays

Effective width of source

Gamma-rays, like X-rays, cover a range of wavelengths in the electromagnetic spectrum roughly from 5×10^{-9} to 5×10^{-11}cm and so overlap the X-ray region. All radioisotope sources emitting gamma-rays have a characteristic output spectrum of radiation which cannot be altered in quality, and also have a constant decay rate usually described in terms of a half-life (the time taken to decay to half-intensity). The output spectrum is generally, but not always, in the form of a series of isolated wavelengths (spectrum lines) rather than the continuous spectrum from an X-ray tube, and the energy of these lines is normally measured in electron volts (eV, keV, MeV).

Only a very few gamma-ray sources have a suitable gamma-ray spectrum for industrial radiography (coupled with a useful intensity of output from a small source) and also a practical half-life, so that nearly all industrial gamma-radiography is done with two sources:

· Cobalt-60 (Co^{60}) — half-life 5.3 years
· Iridium-192 (Ir^{192}) — half-life 74 days.

(The numbers 60 and 192 are the mass numbers of the radioisotope.)

Four other radioisotopes have more limited industrial use and will be discussed later:

· Caesium-137 (Cs^{137}) — half-life 30 years
· Caesium-134 (Cs^{134}) — half-life 2.1 years
· Thulium-170 (Tm^{170}) — half-life 127 days
· Ytterbium-169 (Yb^{169}) — half-life 31 days.

All these gamma-rays sources are produced by national atomic energy agencies, supplied in sealed capsules in a series of standard source dimensions from 0.3×0.3mm to 6×6mm. The activity of a source of gamma-radiation is expressed in curies (Ci) or millicuries (mCi) or in becquerels (Bq), so that a typical industrial radiography source as purchased might be:

· 30Ci Ir^{192} 2×2mm in size, or
· 500mCi Yb^{169} 0.5×0.5mm in size.

The recommended SI unit for source strength is the becquerel (Bq)

$$1Ci = 3.7 \times 10^{10} Bq$$

but is not yet widely used.

It is usual to describe the radiation as Ir^{192} gamma-rays rather than specify the wavelengths of the radiation, and this radiation quality will, for all practical purposes, be the same for all Ir^{192} sources irrespective of source strength, age or physical size.

It is important to realise that, whereas an X-ray set, being an electrical machine, can be switched on and off, a gamma-ray source is never 'off'. The source is held in a thick walled container which is opened and closed, but the source itself is emitting radiation continuously in all directions. When the container is closed this radiation is absorbed in the container walls. Although a gamma-ray source decays in strength with time this decay is exponential and never reaches zero output.

Properties of X- and gamma-rays

The main properties of X-rays and gamma-rays which are of significance in industrial radiography are:

(a)   they are invisible;

(b)   they travel along straight lines;

(c)   they cannot be deflected by lenses, prisms or mirrors (this is not strictly physically correct, but the possible effects of reflection and refraction are so slight as to be practically useless for radiography);

(d)   they pass through all materials, the degree of penetration depending upon the material and the radiation energy; within the useful practical range of X-ray energies the higher the energy the greater the penetration;

(e)   they can damage or destroy living cells, and so are potentially hazardous;

(f)   they are scattered by matter; that is, some secondary X-radiation, which travels in a different direction from the primary beam, can be generated;

(g)   they travel at the speed of light.

Equipment

X-ray apparatus

An X-ray set consists of the X-ray tube, a high tension generator to produce the required kilovoltage, a control panel, a support for the tube and usually a cooling system for the tube (Fig.7).

The conventional bipolar tube, already described in Fig.5, emits a conical beam of X-rays of half-angle about 30° in one direction, and unipolar forms of this tube, where the anode is at earth potential and the cathode at a negative

7 140kV X-ray set showing tube mounted on positioning traverse. Inset: control unit.

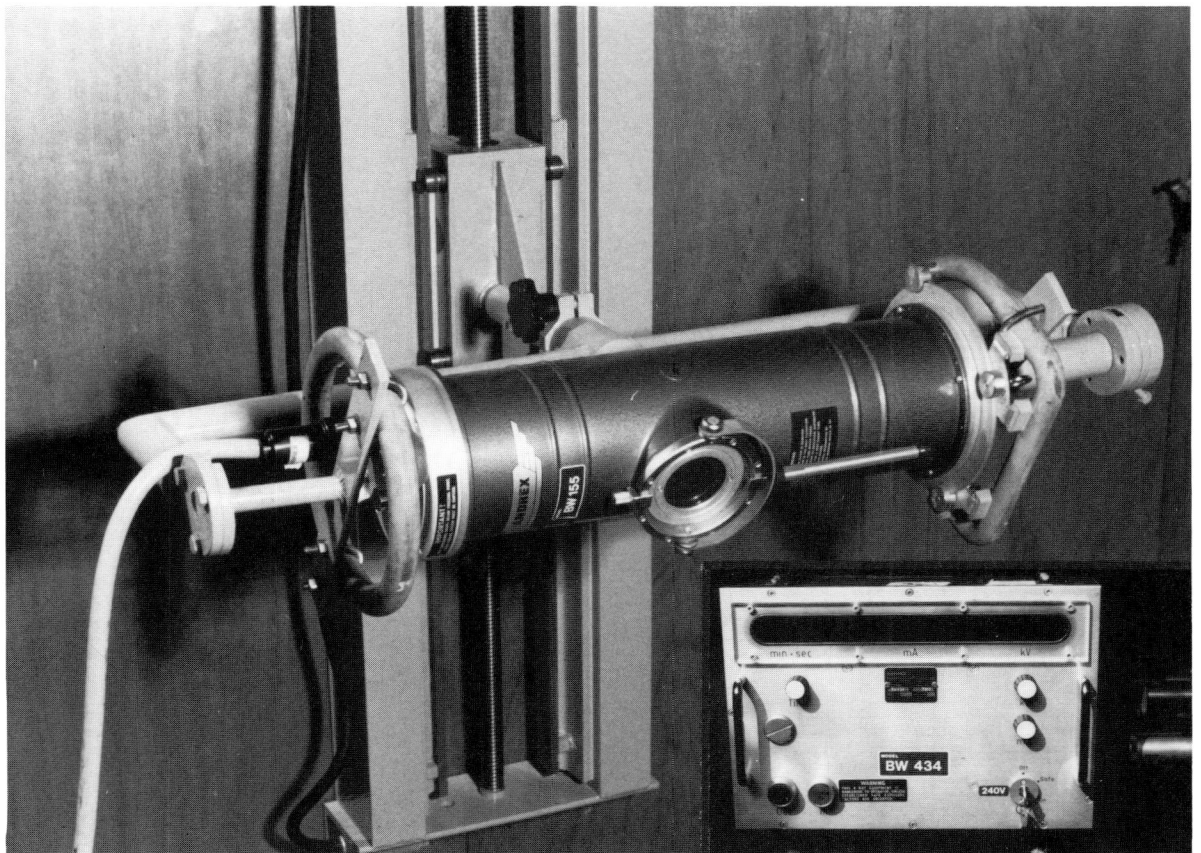

8 Rod anode tube for
360° panoramic X-ray
beam.

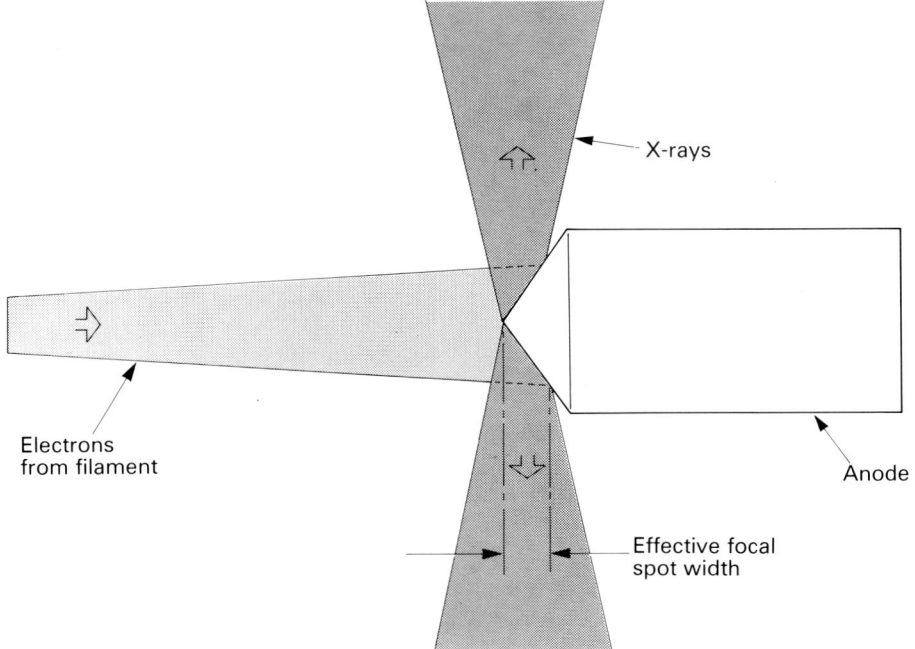

kilovoltage, are also common. Long (rod) anode tubes of this form are also available, and some produce a full 360° panoramic beam (Fig.8), which is very useful for the radiography of large circumferential welds.

The high voltage generator may consist of either a simple step-up transformer with an X-ray tube acting as a rectifier; or a transformer/ rectifier/condenser circuit which can produce a reasonably smooth DC voltage, or a high-frequency Cockcroft-Walton multiplier and inverter which can produce a very uniform high voltage. For voltages below 450kV, the full voltage is usually generated and applied across the X-ray tube.

For higher X-ray energies, other means of accelerating the electrons to a very high velocity are used. In the betatron the electrons are caused to travel in a circular orbit in a doughnut-shaped tube between the poles of a large electromagnet; the electrons are given an additional pulse of energy on each orbit, which causes them to increase in speed, and after a certain number of orbits they are deflected on to a target to generate X-rays. Betatrons have been built to generate 10-31MV X-rays.

Linear electron accelerators (commonly called 'linacs') accelerate pulses of electrons by causing them to 'ride' on a beam of microwaves travelling in a straight line along a corrugated waveguide tube; the electrons increase in energy and velocity along the waveguide and strike a target at the end of the waveguide to produce a beam of X-rays. Usually the X-rays transmitted through the target are used for radiography. Linacs have been built for the generation of 1MV up to 25MV X-rays, and their special characteristic is that they can produce very large outputs of X-rays. Both betatrons and linacs are used for the radiography of thick ferrous weld joints in the range 100—500mm steel.

There are other forms of high energy X-ray generator (van de Graaff, resonance transformer, microtron) which operate in the range 1MV upwards.

In conventional X-ray equipment (less than 400kV), the tube and generator may be connected by HT cable or may be built into a single tank with oil or gas insulation. The type of generator used depends upon the X-ray output desired and whether output, or weight and portability, are the more important factors. Many X-ray sets, up to 300kV, have been designed deliberately in the form of a long cylinder capable of being put through a standard ship's manhole.

Most X-ray equipment up to 400kV is designed to operate over a range of voltages from about 40kV up to the designed maximum voltage, and the control box also usually has a built-in timer, overload trips, safety interlocks, etc. Higher energy X-ray equipment is usually designed for one output voltage only. The mounting arrangements for the X-ray tube or set are of very considerable importance, especially if the set is used in a radiation-shielded room in which the work is brought to the X-ray set. There are no standard designs, but one in which the tube head is moved vertically, horizontally in two directions and has a tilting motion (all by remote control), and is supported on an overhead gantry, is the ideal mounting for general work. Much of the time involved in X-radiography is in setting up, and a well-designed mounting can greatly increase the rate of production of radiographs. Whatever tube movements are available, the system must be rigid.

The design of X-ray laboratories should be based on the specific applications envisaged. If the X-ray beam can be confined to one direction, say downwards, the side walls need provide protection only against scattered radiation — not the full direct beam — and so can be much thinner. The commonest materials for the walls are concrete or lead-plywood (a layer of lead as the centre ply). Hydraulically operated doors are usually used, which interlock with the X-ray set controls. The size of these depends upon the size of specimens to be taken into the X-ray room. If the laboratory is roofless to enable specimens to be lifted in by overhead crane, there can be difficult protection problems with air-scattered radiation ('sky-shine'). Advice on laboratory design and monitoring facilities can be provided by the National Radiological Protection Board.

For much site work, the X-ray tube support is improvised with wood blocks, ropework, etc., but for certain applications readymade forms of mounting — for example, tripods, trolleys, crawlers and crane arms — are used. This remains, however, a somewhat neglected design area. Figure 9 shows a close up of a crawler. Figure 23 shows an X-ray crawler in use for pipeline radiography.

The requirements in terms of X-ray kilovoltage for different specimen thicknesses are given in Table 1. These are approximate values for guidance only on the general requirement for high quality radiography, which will be discussed further in the paragraphs on technique.

Below are some general comments on X-radiography for different weld thicknesses:

(a)   for use on light alloy welds and on steel welds up to about 30mm X-ray equipment is relatively cheap, portable, and reliable;

(b)   for use on steel up to 80mm the equipment is large, more expensive, but still transportable;

(c)   for use on steel thicker than 100mm special equipments such as linacs are necessary; these are rarely portable, are very expensive and the protected laboratories in which they have to be used are also very costly;

(d)   radiography can be applied to any material (up to a maximum of 500mm steel or equivalent) given suitable equipment.

9 X-ray pipeline crawler
(see also Fig.23).

Table 1. *Maximum thicknesses for various radiographic
equipments*

	Maximum thickness which can be examined, mm	
X-rays, kV	Steel	Aluminium
50	1	12
100	18	60
150	30	80
200	50	100
300	80	150
400	100	200
2000 (2MV)	200	500
8000 (8MV)	400	1400
25000 (25MV)	500	1400

Gamma-ray equipment

A gamma-ray source emits radiation continuously. To make it safe to handle,
therefore, the source is shielded by a thick walled container when not in use. A
radiograph is made by opening an aperture in the container or pushing out the
gamma-ray source on a flexible cable once the container has been put in position.

There are three basic designs of source container (sometimes called
'gamma-ray camera'), shown schematically in Fig.10-13. The most efficient
weight-for-safety is obtained by using very dense metal for construction, and
most containers are made of lead, depleted uranium or tungsten alloy. Lead
containers, however, may not meet all the radioactive material transport
regulations in some countries.

The Teleflex-cable type of container (Fig.10 and 11) is becoming
increasingly popular because it is not necessary to position the container

10 Gamma-ray source container: Teleflex type with collimator for 360° panoramic beam. NB: In practice the drive is extended to a safe distance (e.g. 15m).

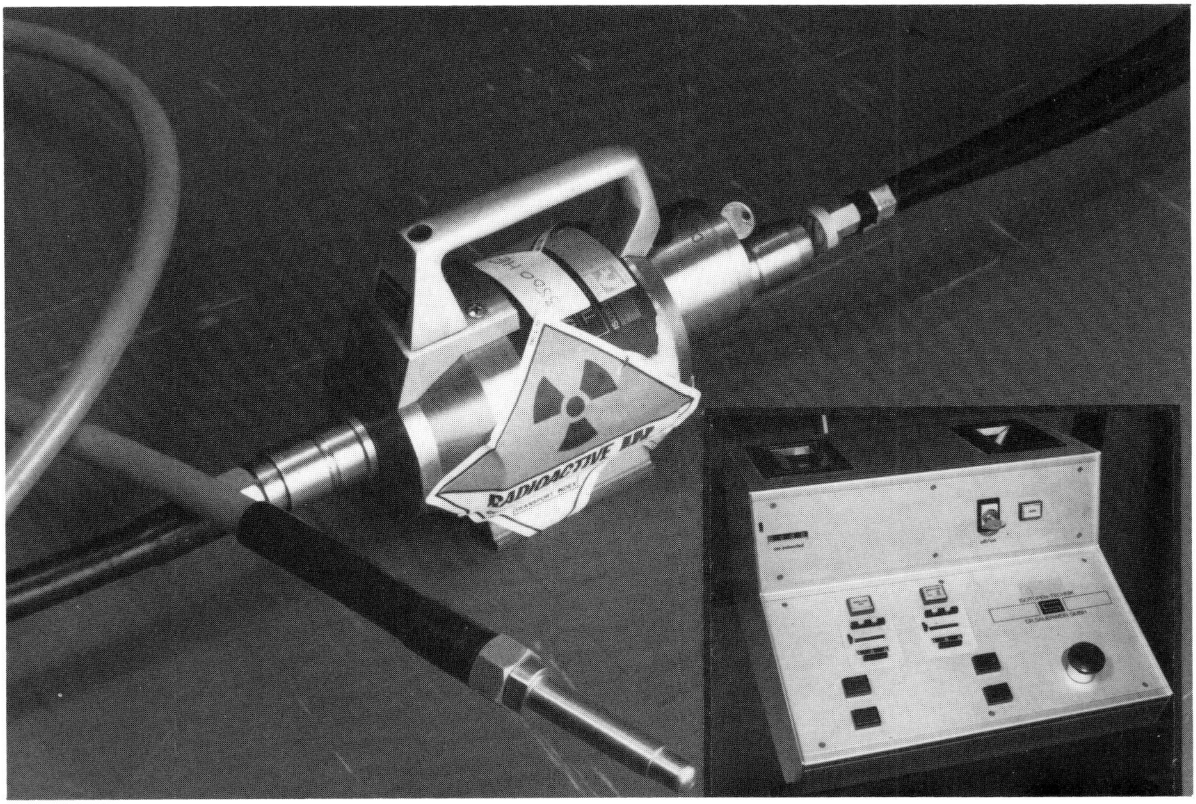

11 Gamma-ray source container for 20 Ci of Ir^{192} with teleflex cable to convey source to exposure head. Inset: electrically driven remote control unit which can be interlocked with enclosure door.

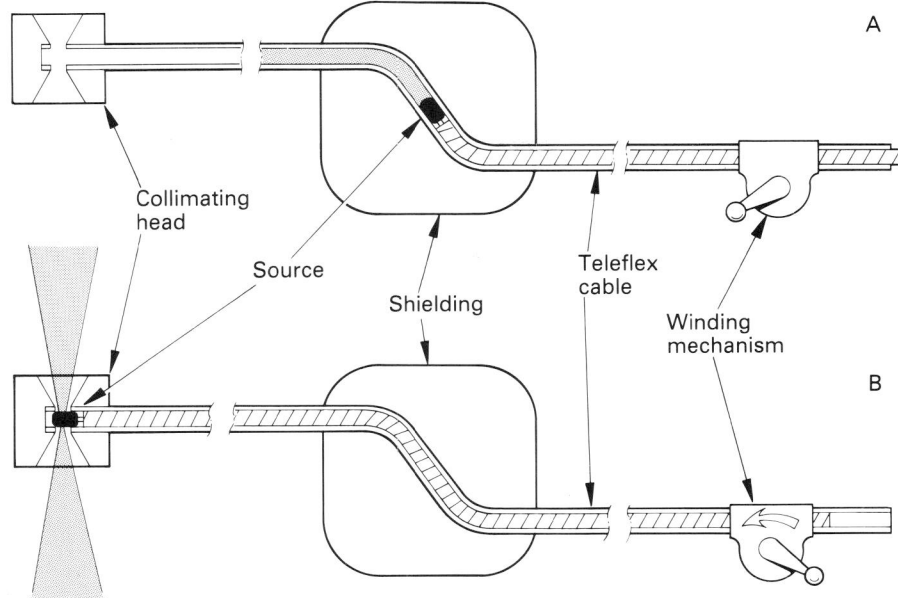

Collimating head

Source

Shielding

Teleflex cable

Winding mechanism

A

B

accurately: a small collimating head is positioned where the source is required to take the radiograph, the flexible extension tube fitted into this, and the main source container can be 10–15m away and remain on a small trolley or vehicle. The collimator is a protective shield to reduce the size of the radiation field. It reduces the scattered radiation reaching the film and the radiation hazards to personnel. The drive mechanism to move the source from its main container into the exposure head is normally 10–15m from the container, so that the operator can be 20–30m from the source in its exposing position. The drive mechanism can be hand-driven, or may be electrical with timing and delay-operation circuitry. Important design points are the source attachment to the cable, the resistance of the extension tubes to external damage and the time of transfer of the source. A container for 30 curies (Ci) of Ir^{192} will weigh about 10kg and one for 30Ci of Co^{60} about 150kg.

With source containers of the types shown in Fig. 12 and 13 a compromise is usually necessary between weight and safe-handling time, and some containers are deliberately designed for limited handling times so as to be easily portable. It is good practice in such cases to have a dummy container for setting up an exposure, and to exchange this for the loaded source container only when the exposure is completely set up.

12 Gamma-ray source container with removable cap.

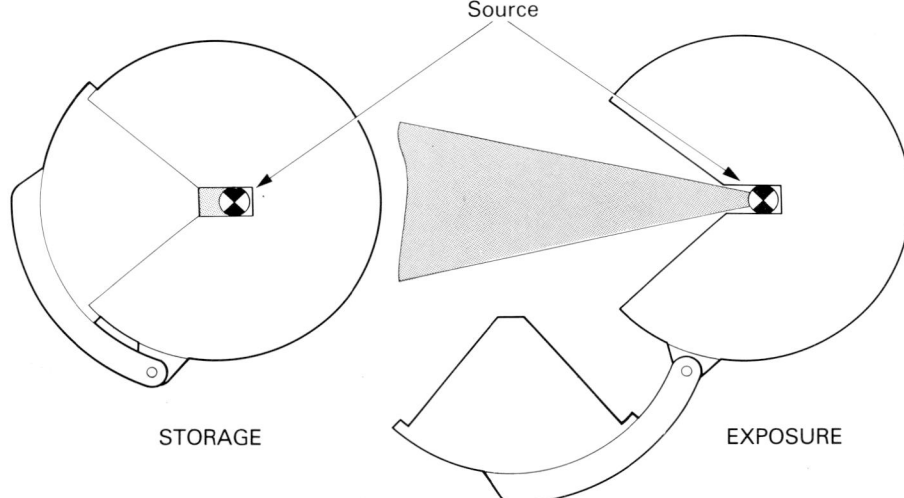

13 Gamma-ray source container with rotating core.

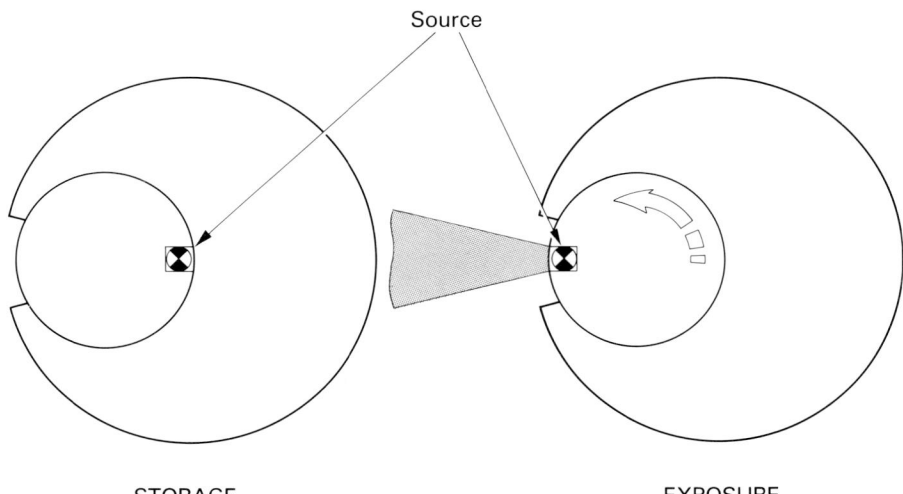

Because the radiation emitted by a gamma-ray source cannot be varied in quality there is a limited range of material thickness for which each source can produce satisfactory radiographs (see Table 2).

Table 2. *Thickness ranges in steel for which gamma-ray sources will produce satisfactory results*

Source	Thickness (approx.), mm
Iridium—192	10—75
Cobalt—60	50—150
Caesium—137	25—100
Caesium—134	35—100
Thulium—170	5—10
Ytterbium—169	2—10

For thicknesses above this range exposure times become too long to be practical, and for thicknesses below it the radiographic image is usually too poor to be acceptable for high quality applications (see also the section on 'Image quality' below). Again, these thickness ranges are for guidance only; they are not strict limits. Figure 14 shows an operator setting up for the exposure for gamma-radiography of a thick-walled container using a Co^{60} gamma-ray source. No gamma-ray sources are generally available which will produce good radiographs of welds in light alloy or plastics materials.

14 Radiography of thick walled vessel using large (1000 Ci) Co^{60} gamma-ray source: setting up for exposure in a shielded bay.

Generally, gamma-radiographs are not as good as the best X-radiographs of the same specimen, particularly on thin welds, but the equipment is cheaper than X-ray equipment, more easily portable and can be independent of electrical and water supplies. Exposure times are usually from about ten minutes up to several hours, compared with typical times of 1—10min in X-radiography.

Gamma-ray sources are normally replaced after they have decayed for one to two half-lives and a replacement source typically costs £100—£500, depending upon its strength and transport costs.

Film and intensifying screens

The radiographic image is obtained on special photographic film which has a thick coating of emulsion on both sides of the cellulose acetate or polyester base. The emulsions are silver halide in gelatin, as in ordinary photographic film. For radiography, the film is normally used with an intensifying screen pressed into good contact on either side of the film. For very low X-ray voltages (below 40kV) no intensifying screens are necessary. With very high energy X-rays (greater than 4MeV) and Co^{60} gamma-rays special screen materials may be preferable, but for all other radiations the intensifying screens used are very thin lead foil (0.05—0.3mm) usually mounted on card. Table 3 shows the appropriate screen thicknesses for different radiations. Lead intensifying screens absorb some of the incident X-rays and emit electrons into the film emulsion: they enable exposure times to be shortened by a factor of between two and six, with only a very small, almost unnoticeable, loss in image quality.

Table 3. *Metal intensifying screens*

Radiation	Screen material	Screen thickness, mm	
		Front	Back (min.)
X-rays			
40—120kV	Lead	None	0.1
120—250kV	Lead	0.03—0.05	0.1
250—400kV	Lead	0.05—0.15	0.1
1MV	Lead	1.5—2.0	1.0
5—10MV	Copper	1.5—2.0	1.5
15—31MV	Tantalum	1.0—1.5	None
Gamma-rays			
Ir^{192}	Lead	0.05—0.15	0.15
Co^{60}	Copper	0.5—2.0	0.5
	Steel	0.5—2.0	0.5

Salt intensifying screens (fluorescent screens), usually consisting of a layer of calcium tungstate, absorb X-rays and emit ultraviolet (UV) light. They have a very large intensifying effect (between ×20 and ×100) and are extensively used in medical radiography. Used with special screen-type film, they can produce large intensifying factors but are rarely used for industrial radiography nowadays because they give a grainy, poor quality image which can obscure fine detail.

The film plus intensifying screens must be held for exposure in a light-tight container. The conventional X-ray cassette has a front of 1mm thick aluminium or plastic, and a pressure pad to squeeze the film and screens tightly together. Cassettes usually have a lead backing to reduce secondary X-ray back scatter (see below). Rigid curved cassettes are marketed, but plastic double envelopes are more commonly used when the film needs to be curved. Poor film/screen contact is a common and serious fault when such plastic cassettes are used and needs to be guarded against; at one time vacuum cassettes were commonly used, but seem now to be rare. For very low voltage X-rays ($< 50\text{kV}$) special thin-fronted cassettes are necessary.

Film characteristics in relation to image quality

The characteristics of the film are a most important factor in the production of high quality radiographs.

The image on a film, after exposure and processing, is measured by its blackening, or photographic density, D. When the film is placed on an illuminated screen for inspection the image is made up of many areas of different brightness, dependent on the local density of the emulsion. Photographic density is defined as:

$$D = \log_{10}\left[\frac{\text{intensity of light incident on film}}{\text{intensity of light transmitted through film}}\right]$$

Thus, if one-tenth of the light incident is transmitted, the film has a density of 1.0 ($\log_{10} 10 = 1$) and if $1/100$ is transmitted the film density is 2.0. Film densities are measured with electronic or optical instruments called densitometers, or can be estimated approximately by visual comparison with a calibrated film density step-wedge.

If a series of different exposures is made on a film, either as a series of steps or as separate pieces of film, and the densities are measured after careful processing, these densities can be plotted against the exposures given. When the density, D, is plotted against $\log_{10}$(exposure) this curve is known as a characteristic curve (Fig.15); that is, it is characteristic of the particular type of film for the particular development process used. 'Exposure' in this context means the product of radiation intensity and time, and is nearly always referred to as:

(milliamperes×min)   mA×min = (X-rays), or
(curie×hr)   Ci×hr   = (gamma-rays), or
roentgens   R.

With exposures using metal foil screens, or no intensifying screens, the absolute exposure time does not matter: thus, 1mA for 10min will give exactly the same result as 100mA for 6sec. This reciprocity does not apply to the use of

15 Typical characteristic curve for radiographic film showing film contrast measured by gradient G_D (the slope of the tangent, shown at density 2.0).

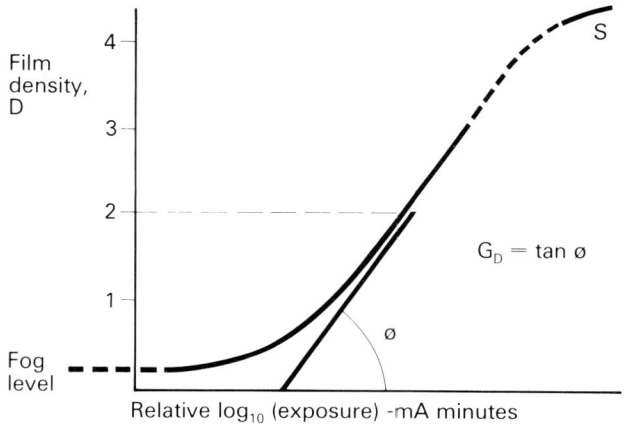

salt (fluorescent) intensifying screens, so the following remarks on film characteristics are not generally applicable to salt screen exposures.

Films for use with metal screens are usually described as either 'direct-type' or 'non-screen' types, and most manufacturers make at least three varieties, i.e. very fast, medium speed and slow. Film designed for use with salt intensifying screens can be used with metal screens, but this generally gives lower contrast results.

The $D/\log_{10}E$ curve for all direct-type films (Fig.15) shows some common features. A low density is obtained with no exposure, i.e. the fog level (usually 0.15–0.25): the curve then has a toe region which merges into an approximately straight region and gradually increases in slope. At extremely high densities, usually well beyond what can easily be measured, there is a shoulder to the curve and a maximum density of the order of 12–14.

The gradient (slope) of the $D/\log_{10}E$ curve (Fig.15) is called the *film contrast* at that density, G_D, and is a most important parameter in radiography. If this gradient is plotted against film density (Fig.16), it can be seen that the film contrast continues to increase with film density, up to very high densities, and does not have a peak value within the commonly used film density range (densities 1.0–3.5).

The importance of film contrast is illustrated in Fig.17. If a thickness change on a specimen is radiographed it produces a certain exposure change on the $\log_{10}E$ axis. If two films having different characteristic curves (A and B) are

16 Typical contrast (gradient) against density curve for fine-grain X-ray film.

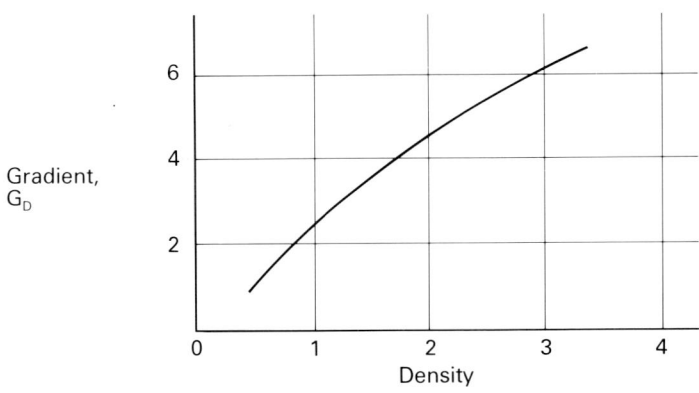

17 Significance of the shape of the characteristic curve for image contrast on a radiograph of a small step in a plate specimen. The film with the steeper gradient (A), although slower, produces a greater density difference for the same thickness change.

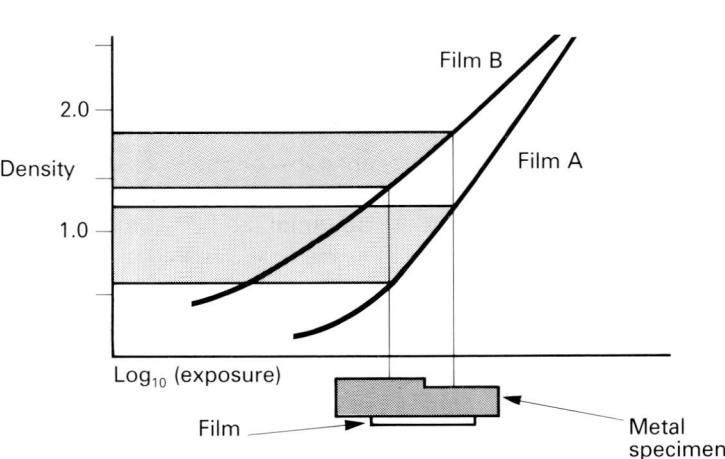

18 Typical characteristic curves for radiographic films of different speeds:
A   high speed;
B   medium grain, medium speed;
C   fine grain, slow;
D   very fine grain, very slow.

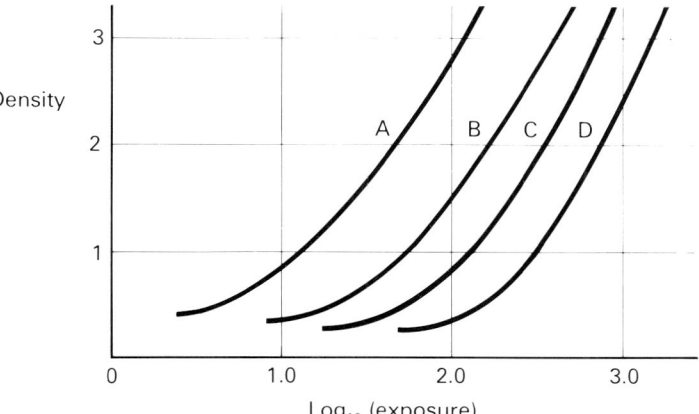

considered, it is obvious that the film with the steeper gradient will produce a greater density difference on the radiograph for the same thickness change in the specimen; that is, the image will have greater contrast.

The shapes of the $D/\log_{10}E$ curve and the gradient/density curve depend on the film processing conditions and the type of film used, but are almost independent of the radiation energy within the usual range of X- and gamma-rays used in industrial radiography.

If the $\log_{10}E$ axis is given absolute units of exposure, such as roentgens, the position of the curve along the axis is a measure of the film speed: the further the film curve is to the left, the smaller the exposure needed to obtain a chosen film density. It has been recommended that a film speed be defined as the reciprocal of the number of roentgens to produce a density of 2.0, but this is not yet widely applied. Figure 18 shows a family of curves for the different direct-type films of one manufacturer.

All radiographs give the appearance of a grainy image, easily visible with a low power magnifying glass. The causes of this graininess are complex and rarely quantified. The granular or mottled nature of the image can obscure fine detail in the image, and it is generally assumed that the slower radiographic films, which are often described as 'fine grain' or 'very fine grain', are capable of resolving smaller image detail than the faster coarser grain films, and so are to be preferred for high quality radiography.

Film processing

Conventional film processing is done by having the film in a channel hanger or a clip frame, which suspends the film vertically in a tank of developer, etc. Radiographic developers are designed to give high contrast, and standard development times are conventionally 4-5min at 20°C. With most developers a little more contrast and a slightly greater effective speed can be obtained by developing for up to 8min at 20°C, but the gains are not large. During development the film must be agitated at least once a minute to ensure even development. Considerable care is necessary in processing and drying and in solution maintenance to ensure freedom from spurious marks, streaks, etc., which can cause difficulties in interpretation of the image.

Over the last few years there has been a large increase in the use of automatic processors, in which the use of higher temperatures and special solutions makes possible a dry-to-dry time as short as 5-6min. The automation of film processing also standardises the processing and so improves the reliability of the radiographic process. Most of the automatic processors employ a roller transport system.

Film viewing

The provision of good viewing conditions is extremely important in industrial radiography, particularly in weld inspection. It is quite possible to have information on a radiograph which is not discerned because of too high a film density or too low an illuminator luminance. At low luminances (below 10 cd/m^2) the human eye, even when dark-adapted, cannot see such low contrast images or discern such small details as is possible at higher luminances, and as there is a tendency to increase film densities to utilise the increase in film contrast, it is obvious that a compromise must be made between high film densities for high contrast and the fall-off in ocular performance at these higher densities, if the viewing screen luminance is not increased.

The International Institute of Welding (IIW)[1] has recommended that the luminance of the illuminated radiograph should not be less than 30 cd/m^2 and, whenever possible, 100 cd/m^2. This minimum value requires illuminator luminances of:

For film density 2.0:   3000 cd/m^2
For film density 3.0: 30 000 cd/m^2

The colour of the light should preferably be white and it should be diffuse, but need not be fully diffuse. The radiograph should be examined in a darkened (but preferably not fully black) room, with care being taken that as little light as possible is reflected off the film surface directly towards the observer. A useful guide to stray light is that the luminance of a white card put in place of the film should not exceed 10% of that of the illuminated film. The film edges and any low density areas need to be masked and the film reader should be adapted to the low luminances likely to be encountered.

The establishment of good film viewing conditions is vitally important to successful industrial radiography. Table 4 shows the increase in contrast which is possible if viewing conditions suitable for higher film densities are made available. Most codes of good practice recommend film densities in the range 1.8—2.5. It is seldom that good viewing conditions for film densities greater than 3.5 are available.

Table 4. *Increase in contrast achieved through optimum viewing conditions (i.e. those permitting the use of higher density film)*

Film density	Contrast as a % of value at density 1.5
1.5	100
2.0	130
2.5	163
3.0	185
4.0	230

Image quality

The image obtained on a radiographic film can be assessed in terms of three factors: contrast, definition and graininess, ignoring adventitious factors such as scratches, streaks, dust spots, etc., which can also reduce image quality.

X-ray absorption

When X- or gamma-rays are incident on one side of a specimen, the amount of radiation transmitted to the film on the other side is described by the basic absorption formula:

$$I = I_o e^{-\mu t}, \textit{ alternatively written as } I = I_o[\exp(-\mu t)] \qquad [1]$$

where:

I_o is the intensity of the incident beam;
I is the intensity of the transmitted beam;
μ is the linear absorption coefficient, which is dependent upon the radiation energy and the material of the specimen;
t is the specimen thickness;
exp stands for the exponential (e = 2.718).

This formula is of little practical importance in planning a radiographic technique as it applies strictly only to monoenergetic radiation, which is rarely encountered in practical radiography, but the term μ will be required later, and it is convenient to have it defined at this stage.

The important practical points on absorption and transmission are:

(a) As the kilovoltage is increased, μ decreases — i.e. a greater proportion of the incident radiation is transmitted.
(b) The greater the physical density of the material of the specimen, the larger the value of μ — i.e. the greater the absorption.
(c) When a wide beam of radiation is incident on the specimen, some radiation is absorbed and some is scattered; this scattered radiation can be multiple scattered (Fig.19) so that at the point on the film, P, some of the received radiation travels directly from the source and will form an image of a cavity:

19 Influence of scattered radiation: non-image forming radiation can reach film at point P by multiple scattering but does not contribute to the formation of image of cavity.

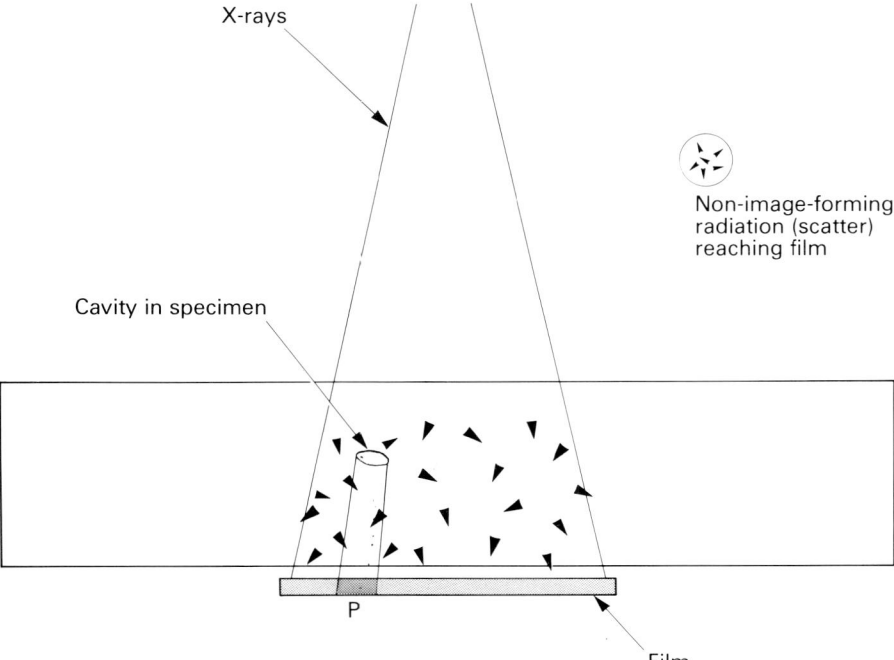

this is the image-forming, direct radiation, I_D. Some radiation also reaches P indirectly by scattering processes, and this is non-image-forming, I_S. The ratio I_S/I_D (called the build-up factor) is usually much larger than unity, in the range 2-30, and depends on the radiation energy and the specimen thickness. Very high energy radiation, in the megavoltage range, produces much less scatter than lower energy X-rays, which is one of the attractions of using megavoltage radiography.

(d) A minimum X-ray kilovoltage has to be chosen which is appropriate to the specimen thickness; this is usually determined from a set of practical exposure curves made for the particular equipment, in terms of an exposure time of practical duration. If too high a kilovoltage is used there is a loss of contrast; if too low a kilovoltage is chosen the exposure times will be impractically long.

Contrast sensitivity

A very convenient measure of radiographic image quality is the density change, ΔD, produced on a radiograph for a small change in thickness of a specimen, Δx, such as a small step. This ratio, $\Delta D/\Delta x$, is called the contrast sensitivity and obviously depends upon both the radiation contrast parameters, μ, I_S/I_D, and the film contrast, G_D. A formula can easily be derived which is very useful in showing the relative importance of the various factors: it is obtained by combining a formula for the shape of the film characteristic curve, and an assumption that the total radiation reaching any point on the film is (I_S+I_D).

$$\frac{\Delta D}{\Delta x} = \frac{0.43 \times \mu \times G_D}{(1 + I_S/I_D)} \qquad [2]$$

If it is assumed that $\Delta D_{min.}$ is a minimum value that the eye can discern under good viewing conditions, Δx can be taken as a fraction of the specimen thickness, x, and $\Delta x/x$ is a measure of thickness sensitivity, S_t.

$$S_t = \frac{\Delta x\ 100}{x} = \frac{2 \cdot 3 \Delta D_{min}(1 + I_S/I_D)\ 100}{\mu G_D x} \qquad [3]$$

For edge images (steps), $\Delta D_{min.}$ is found to be about 0.007, and S_t has values in good radiography of 0.3—0.6%. The importance of this formula is that is shows that S_t can be improved (i.e. made smaller numerically) by:

(a) increasing μ, i.e. using a lower X-ray kilovoltage;
(b) decreasing I_S/I_D (improving the conditions of the set-up to minimise scatter, for example, by filtering the radiation);
(c) increasing the film contrast, G_D (using a higher contrast film, or a higher working density; this is the most important factor, in practice);
(d) making ΔD smaller, i.e. improving the viewing conditions.

Formulae for detail sensitivity (cracks, holes) can be derived from this basic thickness sensitivity formula. Of the factors in this formula, G_D and $\Delta D_{min.}$ are those on which the radiographer can have the most influence. G_D depends on the film type used and the working film density, and $\Delta D_{min.}$ depends on the film viewing conditions; both can vary considerably in value. The other factors in the equation can usually be changed only by small amounts.

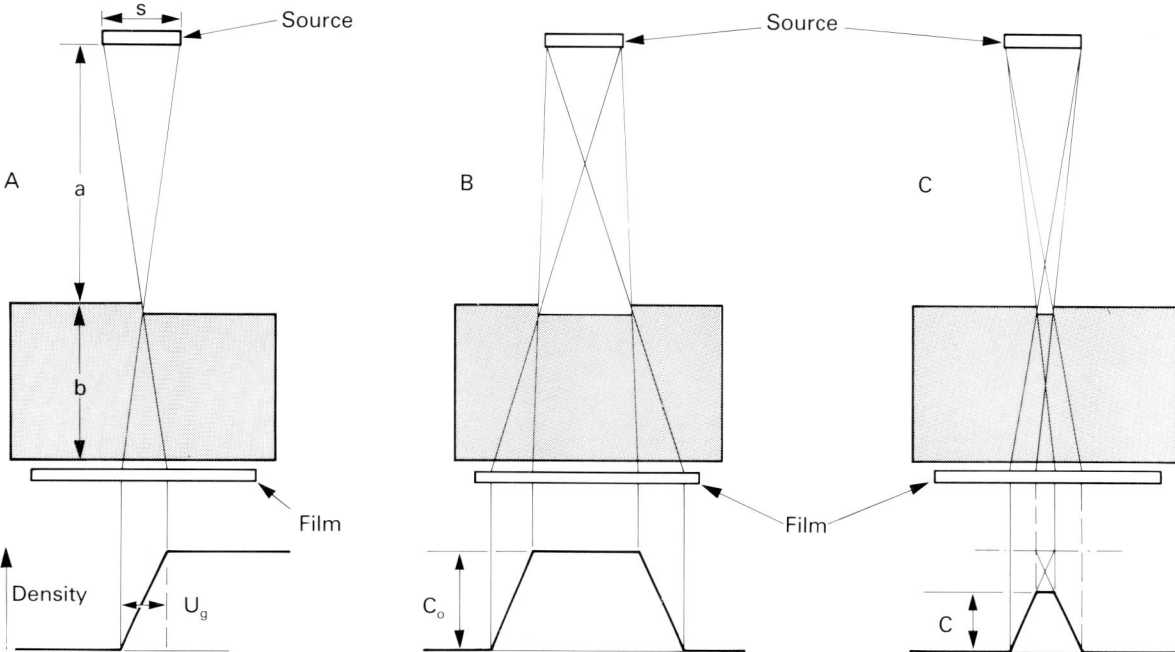

20 Geometric unsharpness, U_g, showing density distribution:
A   in image of small sharp edge formed by source of radiation, s, of large diameter;
B   across image of broad groove;
C   across image of narrow groove: geometric unsharpness causes blurred edges to image and loss of contrast.

U_g is the width of the geometric unsharpness (usually measured in mm); C_o is the image contrast if there is no unsharpness; C is the actual image contrast obtained in the presence of unsharpness. NB: The source is drawn with a large diameter, for clarity.

Image definition

Because the image is formed by radiation travelling in straight lines from source to film, and the source is always a small area and not a true point source, there is a penumbral effect (Fig.20), usually called the geometric unsharpness, U_g, the width of which is given by the formula:

$$U_g = \frac{sb}{a} \tag{4}$$

If a large detail in the specimen is considered which has sharp edges (Fig.20b), this simply means that the edges of the image are blurred and the width of the blurring is U_g. If, however, the detail being imaged is small and U_g is larger, there will be a second effect, a loss of image contrast (C compared with C_o) as well as edge blurring (Fig.20c). The facts of geometric unsharpness are quite simple:

(a)   it increases as the detail being imaged is further from the film, i.e. it is likely to be a more serious problem with thicker specimens;
(b)   it can be decreased by using a larger source-to-film distance (sfd);
(c)   it can be decreased by using a smaller source diameter;
(d)   it is less important if the detail being imaged is close to the film side of the specimen.

In practice, however, X-ray tube focal spots are not uniform emitters, nor sharp-edged, and Fig.20a is a considerable simplification of the practical case.

Inherent (film) unsharpness.   Even if the geometric unsharpness is made negligibly small the image on a film taken with X-rays is not sharp, because secondary electrons are released by absorption of X-rays in the emulsion and these spread sideways, so affecting a small volume of silver halide crystals around the original point of absorption of the X-rays. This means that the image of each point is a disc instead of a point, and it is convenient to measure this as a blurring, analogous to geometric unsharpness. The magnitude of this film unsharpness is very small with low energy X-rays, but is an important factor with high energy X-rays and gamma-rays (Table 5).

Table 5. *Typical values of U_f, film unsharpness*

Radiation	U_f mm
100kV X-rays	0.05
400kV	0.16
8MV	0.50
Ir^{192} gamma-rays	0.17
Co^{60} gamma-rays	0.4

To obtain the total unsharpness on a radiograph U_g and U_f must be combined, and there is no point in making one much smaller than the other. A useful rule in the determination of a technique is to make U_g equal to U_f, or for critical techniques $U_g = \frac{1}{2}U_f$, and then, from the calculated value of U_g for a known weld thickness and radiation source diameter, equation [4] is used to calculate the minimum source-to-film distance (sfd) to be used.

Film graininess.   Because the image on a radiograph is made up of grains of silver and there are only a few hundred per square millimetre, the image has a granular appearance, which can obscure fine detail.

With low energy X-rays, because thin weldments are being examined and small detail sought, it is normal practice to use fine grain or very fine grain film. With very high energy X-rays, because of the high X-ray outputs of most equipments, fine grain or very fine grain films can be used without any problems of unacceptably long exposure times. With the intermediate energies a compromise has to be made between the better detail recording of fine grain films and the necessarily longer exposure times. Most specifications for weld radiography ask for fine grain or medium grain films to be used: very fine grain films are generally considered to be too slow, and high speed films too grainy.

Other technique factors

With high energy X-rays (> 4MeV) and with Co^{60} gamma-rays, intensifying screens of copper or steel will provide better quality radiographs than thin lead screens; also, thicker screens are advantageous. Some radiographic films, however, particularly film strip for wrapping around circumferential seams, are available ready-packed with lead screens in the film envelope, to be used once and then discarded.

Not all weldments are of uniform thickness nor is it always practical to take separate radiographs for each thickness. If a weld of varying thickness is radiographed on one film with a single exposure, there is a danger that the thinner parts will be recorded at a very high density which cannot be viewed satisfactorily, and that the thicker parts will be recorded at a low density where the film contrast is low and the detail sensitivity is poor. There are various procedures with filtered higher energy radiation, double film techniques and variations in developing time, to overcome these limitations.

Summary of technique parameters

Unquestionably, the most important factor is the film contrast, which is controlled by the choice of film type, working film density and correct film processing. Secondly, a source-to-film distance (sfd) must be chosen which, in terms of the source diameter and specimen thickness, gives a reasonable compromise value of U_g in relation to U_f. Thirdly, provision of adequate film viewing conditions is essential. Much less critical are such factors as X-ray kilovoltage and intensifying screen thicknesses. If, however, gamma-rays are used instead of X-rays, this will nearly always lead to a loss of image quality except on thick weldments.

Image quality indicators (IQIs)

Because there are so many variables in a radiographic technique (choice of kV, sfd, film type, screens, development, etc.), it is necessary to have some criterion from which it can be judged whether a radiograph has been obtained using a satisfactory technique. This is often done by placing some object of known form and size on the specimen so that its image appears on the film. These devices were originally called 'penetrameters' but now, in Europe at least, the preferred name is 'image quality indicator' (IQI).

There is no single standard pattern for IQIs, but BS 3971: 1985[2] describes three basic designs of varying degrees of complexity, each with its own special merits. One of these is widely used in Germany, Austria and the Nordic countries, the second in France, and the third (relatively new) only in the CEGB in the UK. The USA uses a quite different design and still uses the name 'penetrameter'.

21 Three commonly-used types of IQI:
A wire type;
B step/hole type;
C American ASTM plaque type.

The first design (Fig.21a), known as the wire type, consists of a series of round wires of different diameters of the same material as the specimen to be radiographed. Each wire is 30mm long, and the IQI is placed on the specimen on the side remote from the film; most codes of practice specify that it be placed

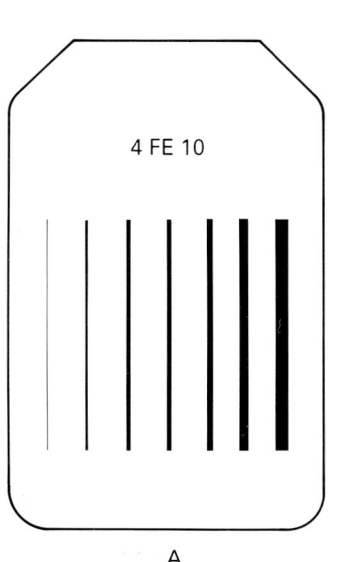

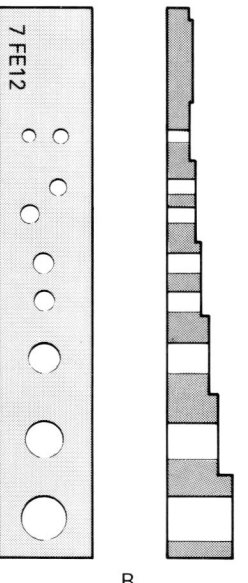

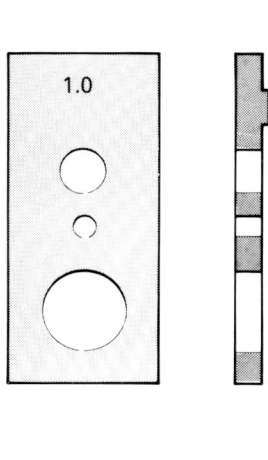

A   B   C

across the weld, near one end of the radiograph with the thinnest wire away from the centre of the film. The image is examined on the processed film and the thinnest wire, the image of which can be discerned, is selected. The diameter of this wire, expressed as a percentage of the weld thickness is the 'IQI (wire) sensitivity' or sometimes 'DIN wire sensitivity', and on a good quality radiograph will be between 1.0 and 2.0%, depending upon the weld thickness.

BS 3971: 1985 includes tables of acceptable IQI sensitivity values as guides to what ought to be obtained with satisfactory techniques on different steel thicknesses. Because the IQI sensitivity is expressed as a percentage, the smaller the numerical value, the better the sensitivity, i.e. a smaller wire is detected, and it is assumed, therefore, that smaller weld flaws will be detectable. It seems obvious that there must be some relationship between IQI sensitivity and flaw sensitivity, but the relationship is complex, and it certainly is not true that, say, 2% IQI (wire) sensitivity means that all flaws occupying 2% of the thickness will be detected.

The second type of IQI — the step/hole type — consists of a stepped wedge (Fig.21b) with a drilled hole in each step of hole diameter equal to step thickness. (This IQI is also often known as the AFNOR pattern.) Again, sensitivity is measured by the size of the smallest discernible hole when the IQI is placed on the side of the specimen remote from the film; this is designated IQI step/hole sensitivity. The American (ASTM) design of IQI is somewhat similar in principle in that it uses drilled holes but, instead of a stepped wedge, a uniform thickness plaque with three holes of different diameters is used (Fig.21c), and the methods of calculating percentage sensitivities are more complex.

It is important to realise that 2% sensitivity (or any other value) using a wire type IQI is not the same standard as 2% on a step/hole type or on an American plaque type. Also with one type of IQI 2% may represent a very good radiograph; with another, 2% sensitivity may be a relatively poor quality radiograph.

The fourth type of IQI (BS 3971: 1985; type IIIA) is entirely different. It consists of pairs of wires of very dense metal, each pair of wires being separated by a space equal to the wire diameter. In the smallest element of the IQI the wires are 0.1mm diameter and the largest 0.80mm diameter. There is also a model B for high energy X-rays and gamma-rays in which pairs of bars are used instead of circular cross-section wires. On the radiograph the images of the finer pairs of wires merge because of image unsharpness. By deciding on the finest pair which can be seen as two separate wires, a measure of the acting total unsharpness is obtained. This pattern of IQI is more expensive to construct than the normal wire type described above, and is as yet not widely used.

Image quality indicators have a limited value in that the reading obtained is not very sensitive to changes in radiographic technique, and as a personal judgement has to be made of when an image is 'just discernible', there can be considerable human error in the IQI values obtained. Nevertheless, most radiographic specifications for weld inspection quote acceptable IQI values which must be attained if the radiograph is to be accepted.

A mathematical relationship can be developed between IQI sensitivity and crack sensitivity, but the detectability of a crack also depends markedly on its opening width and its angle to the radiation beam. A tight crack with a small opening width, or a crack at a considerable angle to the radiation beam, are both unlikely to be detected by radiography.

With volumetric flaws the situation is simpler. A gas cavity is somewhat similar to a drill hole of depth equal to diameter, and there is a direct relationship between IQI (hole) sensitivity and the detection of small gas cavities.

Interpretation

Radiographs are interpreted by examining them by eye when they are placed on a satisfactory film viewing screen. The importance of good film viewing conditions and the difficulties in providing these for film densities higher than 2.0 has already been emphasized, and the levels of screen luminance required have already been given. It is essential to check the near-distance acuity of the eyes of the film reader as much film viewing is done at around 250mm reading distance or less. A low power magnifier is a very useful accessory.

Training and experience are essential for the proper reading and interpretation of radiographs. The reader must have a good knowledge of the various defects which may occur in welds and of their causes. He must also have a knowledge of the welding and radiographic techniques used and be familiar with the characteristic appearances of various flaws and surface imperfections. To differentiate more readily between images due to internal flaws and those due to surface marks, the visual appearance of the weld should be taken into account in interpretation. The correct identification of a flaw can occasionally be difficult, and the difficulties tend to increase with weld thickness because image unsharpness increases with higher energy radiation. Photographic film and intensifying screens can produce spurious marks on the radiographs, the forms of which are detailed in radiographic handbooks. These are, however, rarely confused with weld defects by experienced film readers. In the radiography of light alloy and austenitic steel weldments an effect known as 'diffraction mottling' can produce spurious images caused by the large grain size of the material, and this mottling can sometimes cause difficulties in interpretation.

ISO 6520: 1982[4] describes weld flaws and their radiographic appearance, and various organisations (IIW, ASTM, TWI) have produced sets of reference radiographs which can be used to assist in the identification of flaws. Some of these sets of radiographs illustrate different degrees of severity of certain flaws, such as distributed porosity, and enable terms such as 'slight porosity' and 'heavy porosity' to have a more uniform meaning. Some organisations have used these reference radiographs to specify acceptance standards for certain weld defects for specific applications. This is a somewhat controversial use of reference radiographs, as none of these sets covers a range of weld thicknesses in detail or the variety of radiographic techniques which may be used. Also, the change in appearance of a planar flaw as the radiation beam angle is altered can be quite considerable, and these variations are not covered in existing sets.

Specifications

Most industrial countries have issued national standards in the form of codes of good practice for different applications of weld radiography. In the UK BS 2600: 1983[5] and BS 2910: 1986[6] are the latest (revised) versions of the two basic codes on radiography of welds, but, generally, less detailed codes are issued by ISO and IIW. These codes of practice offer considerable information on good radiographic practice in that they include guidance on correct X-ray kilovoltage, minimum sfd, film density, thicknesses of intensifying screens, processing, etc. for each basic method, but where a choice of methods is possible this choice is generally left to the inspection authority for each individual application.

Radiography of butt welds in pipes

For example, there are four basic possible methods for the radiography of circumferential butt welds in pipes:

· single wall, single image (source inside);
· single wall, single image (source outside);
· double wall, single image;
· double wall, double image.

For each of these, one of three different types of film may be used (medium, fine, or very fine grain). Also, on some thicknesses either X- or gamma-rays may be employed.

Single wall, single image (source inside).   Figure 22 illustrates the single wall, single image technique with the source inside. The technique can be applied to

22 Radiography of circumferential butt welds in pipes I: single wall, single image (source inside):
A   source placed on centreline;
B   source off centre.

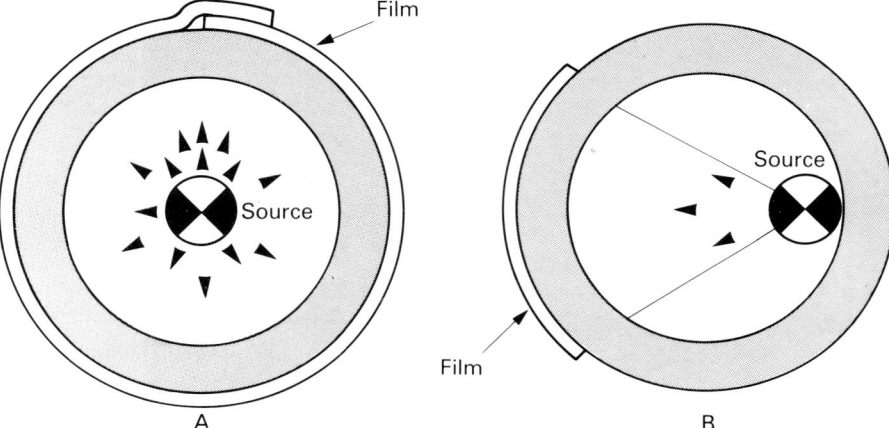

23   X-ray pipeline crawler in use in pipeline radiography (Courtesy BIX Industrial Testing Limited).

24   Radiography of
circumferential butt welds
in pipes II: single wall,
single image (source
outside).

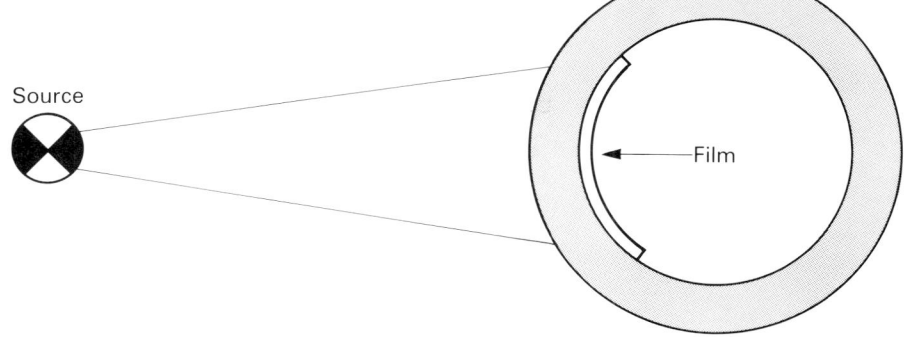

25   Radiography of an oil
feed pipe in a jet engine:
the film is placed inside the
pipe (Courtesy of
Rolls-Royce Ltd).

pipes only where a radiation source can be placed inside. It has the advantage that, if the source can be placed on the centreline and the film wrapped round the pipe, the whole weld may be radiographed in one exposure. This is economical and rapid and is the most commonly used method in the testing of pipelines. Remotely controlled crawlers (Fig.9) can be made to carry a small X-ray set or gamma-ray source along a pipe to the correct location for radiography (Fig.23). The source size and the radiation energy must be matched to the weld dimensions.

Single wall, single image (source outside).   Figure 24 illustrates the single wall, single image technique with the source outside — a technique which requires access to the inside of the pipe to place the film cassette. Between six and fourteen radiographs are needed to cover a 360° weld, depending upon the pipe diameter and thickness. For a fixed pipe, this would be a very time-consuming procedure. Figure 25 shows the technique being used to test a weld in an oil feed pipe in a jet engine.

Double wall, single image.   The next two techniques require no access to the inside of the pipe as the source and the film are both placed outside. However. the radiation used must be able to penetrate two wall thicknesses in order to obtain a radiograph of the weld in one wall, which means that the X-ray kilovoltage must be higher and the flaw sensitivity is poorer than with single wall techniques. The source may be close to the pipe or spaced away (as shown in Fig.26) according to the calculated requirements of the minimum sfd. Figure 27 shows a pipeline weld being examined with an external gamma-ray source using this technique.

Double wall, double image.   If the source is spaced away from the pipe, and offset slightly along the length of the pipe, images of the two sides of the weld are separated (Fig.28). Called the double wall, double image technique, it is, however, only applicable to small diameter pipes (less than 90mm). By using a large sfd, adequate definition can be obtained of the images on both sides of the circumferential weld, which may be covered in two or three exposures.

   With the last two techniques, if there is no access to the inside of the pipe there is no satisfactory position for an IQI, and the method used should be proved on a separate piece of pipe of the same dimensions. Some authorities allow the IQI to be placed on the film surface of the weld, but this is a very doubtful and controversial practice.

   Many applications are restricted to only one of these methods by physical constraints such as access, but there are many instances where a choice has to be made between, say, the method shown in Fig.24 using a gamma-ray source and that in Fig.26 using X-rays through two walls, and at present no existing specifications offer much guidance on this choice. On very small diameter thin-wall steel pipes (1-5mm wall thickness), the situation has been eased by the recent availability of very small diameter Yb[169] gamma-ray sources, used with devices to place the source on the centreline of the weld. The very small source size and the relatively low energy radiation mean that the requirements of image definition and image contrast can be achieved.

26   Radiography of circumferential butt welds in pipes III: double wall, single image (source outside, film outside):
A   source close to pipe;
B   source stood off to produce smaller U_g.

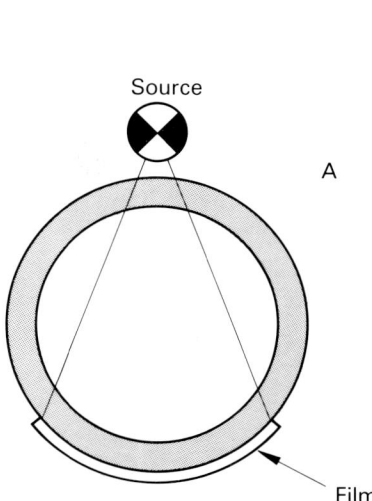

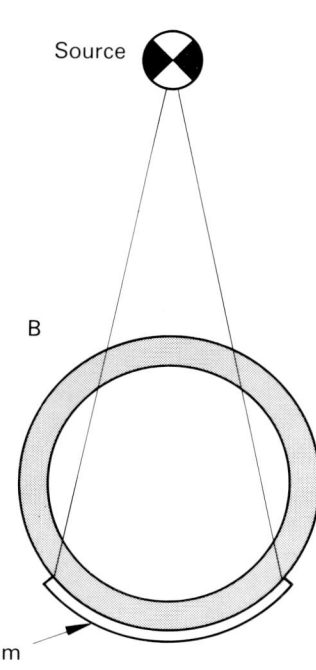

27 Examination of a
pipeline weld with an
external gamma-ray source
using the double
wall/single image
technique.

28 Radiography of
circumferential butt welds
in pipes IV: double
wall/double image
technique for small pipes;
source offset from plane of
weld to produce image
showing both top and
bottom of weld.

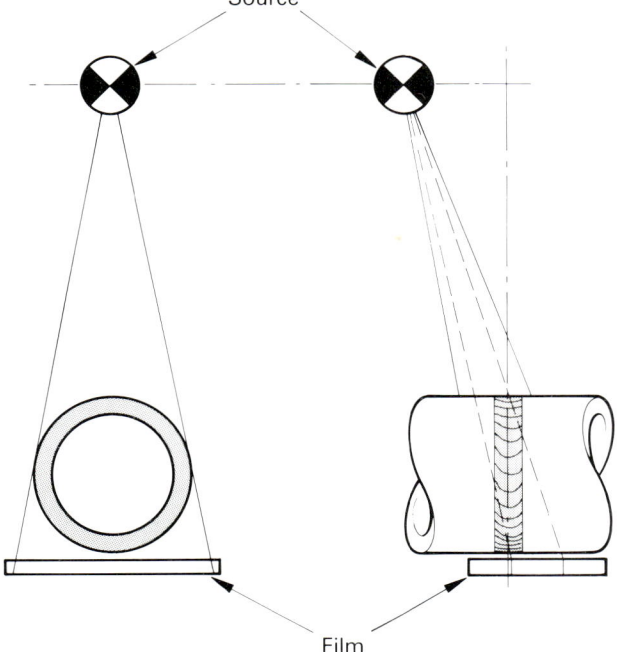

Source

Film

Safety: hazards

All ionising radiations such as X- and gamma-rays are potentially dangerous and
there are statutory requirements on the maximum dose of radiation which an
operator or anyone else may receive in a given period of time. The subject is
complex in that the dose levels are set in terms of the dose which might be

received by the population as a whole, rather than hazard to the individual. In the UK the levels are set by recommendations from the International Committee on Radiological Protection (ICRP) and are overseen by the Health and Safety Executive of the Factory Inspectorate (Ionising Radiations Regulations, 1985).

The SI unit of dose (see below) is the sievert (Sv). The dose limit per year for the whole body, due to external radiation, must not exceed the following values:

- for employees aged 18 years or over: 50mSv
- for trainees aged under 18 years: 15mSv
- for any other person: 5mSv

and there are different limits for individual organs and the eye.

The *UK Approved Code of Practice 1985* requires that the shielding around any radiographic installation shall reduce the instantaneous dose rate outside the shielding to as low as reasonably practicable, and a value of 2.5µSv/hr is taken as the design criterion. The Code of Practice should be consulted for other details. 'Exposed workers' (classified persons) are required to have periodic medical examinations and must carry a dosemeter or radiation monitoring badge. There are special provisions in the *Approved Code of Practice* for site radiography.

Forms of protection

Protection against radiation can be provided in a number of ways of which the basics are distance, time and absorber, or shielding, thickness. For a fixed radiographic installation (Fig.29) the walls of the laboratory are normally thick concrete or lead-lined; access doors are similarly radiation absorbent and are interlocked to the X-ray set controls so that it is impossible to switch on the X-rays while the doors are open. Similar interlocking keys can be fitted to gamma-ray source containers. Often, the beam direction of an X-ray set can be

29 Special shielding enclosure for gamma radiography.

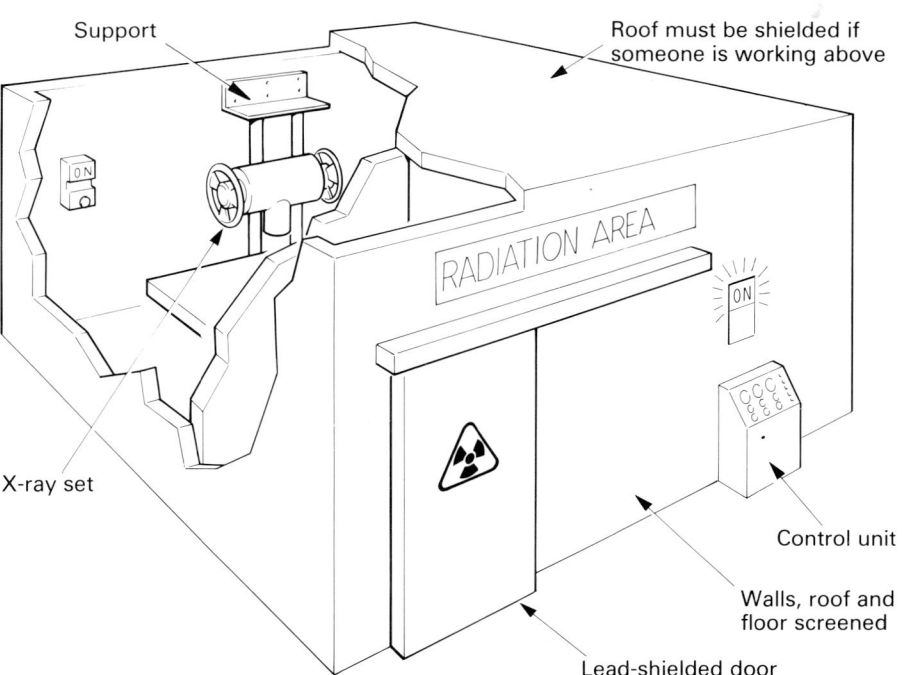

restricted and then not all the walls need have the thickness to protect against the full direct beam; protection against scattered radiation is, however, still necessary. UK national regulations state that warning lights, emergency switches and delay signals (lights, buzzers) must be installed. The cost of a laboratory for radiography rises rapidly with high energy X-rays, but given a properly designed installation no one should ever receive anything approaching the maximum permissible dose of radiation; that is, the hazard can be reduced to zero.

For site work it is sometimes feasible to erect temporary radiation-absorbing screens, but much more emphasis is usually put on using time and distance as the method of protection, together with intelligent positioning of the radiation source. The inverse square law states that at double the distance, the dose rate is one-quarter, etc. and physical barriers with warning notices and lights can be erected at appropriate distances, either after calculation or measurement of the dose rate round the work site (Fig.30). The calculated dose levels should always be rechecked by radiation monitors.

If the radiation source can be positioned so that the direct beam is incident only on to the ground or on to some major absorber, the problems of protection are much simpler, and the barriers need not cover such an extensive area. The protection requirements of gamma-ray source containers are given in BS 5650: 1978[7] (ISO-3999-1977[8]) and depend upon whether the container is one-man portable, or mobile, or fixed. The dose rate must be measured on the container surface at 50mm and at 1m distance. There are several physical tests for gamma-ray source containers and their accessories, and there are separate regulations for public transport of source containers (IAEA, 1973[9]).

30 Site radiography showing safety precautions.

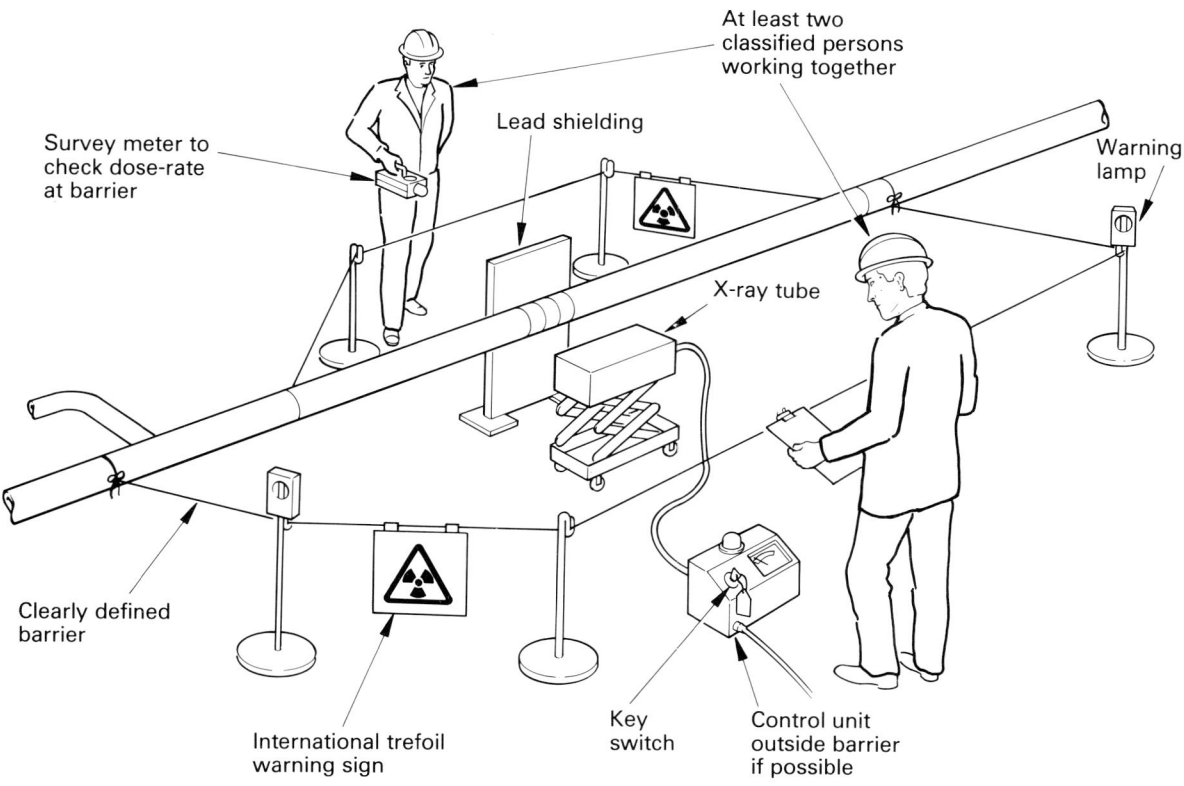

Survey meter to check dose-rate at barrier

At least two classified persons working together

Lead shielding

Warning lamp

X-ray tube

Clearly defined barrier

International trefoil warning sign

Key switch

Control unit outside barrier if possible

Units for measuring radiation dosage

The units in which ionising radiation is measured and in which safety dose levels are specified are quite complicated and were made more so by an ICP recommendation (1975) that new units, compatible with the SI system, be introduced. The new units, however, are not as yet much used in industrial radiography. In what follows, therefore, the old units are given first and the new units second, together with a note on the relation between them.

The old units are roentgens (R), rads and rems. For many years the unit for measuring the quantity or 'exposure dose' of ionising radiation emitted has been the *roentgen* (R), which corresponds to a measure of 83.3 erg/g of energy absorbed in air.[10] The unit of absorbed dose, the *rad*, is the amount of radiation which results in an energy absorption of 100 erg/g in any material (and therefore varies with the material). The *rem* is the unit of absorbed dose which has the same biological effect as one rad of X-radiation (and therefore measures potential damage to human tissue). For practical purposes, however, for the medium energy X-rays used in industrial radiography, the R, rad and rem are roughly *numerically* equivalent.

The new units are coulomb/kilograms, grays and sieverts. In the new units, which were introduced to be more logically compatible with other SI units, the unit of exposure dose is the *coulomb/kilogram* (C/kg), equal to 3876R, but has no separate name. The unit which has replaced the rad as a measure of a given level of radiation absorption in any material is the gray (Gy), equal to 100rad. The unit which has replaced the rem to measure the absorbed dose for a given level of absorption in living tissue is the *sievert* (Sv), equal to 100rem.

Radiation monitoring devices

There are several types of radiation monitor, most of which use an ionisation chamber as detector. Battery-operated, hand held survey monitors (Fig.31) are used to check safety levels of radiation, particularly in site work. Instantaneous personal monitors, carried in pockets or fixed to clothing, give a visible or audible

31 Survey monitor used for routine checking of radiation intensity.

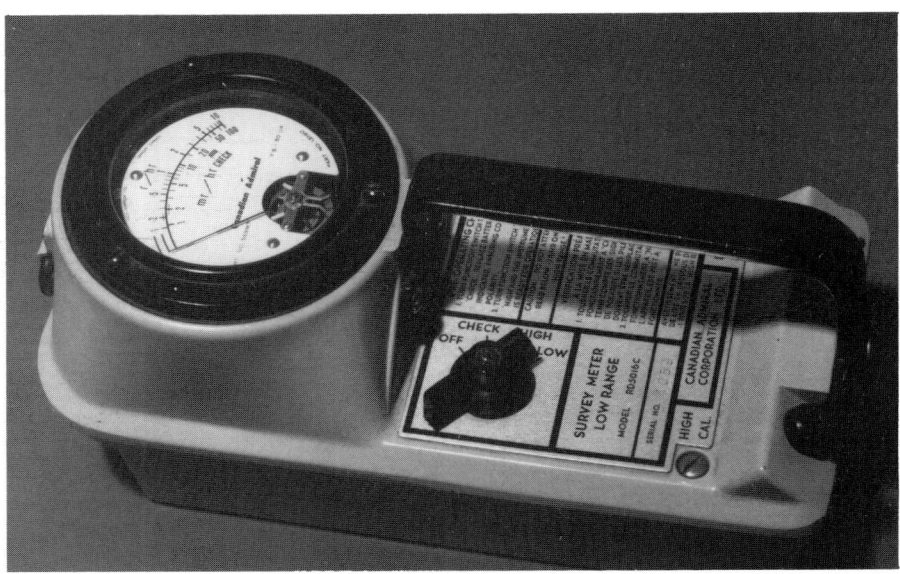

32 Quartz fibre electrometer (QFE) giving a direct readout of accumulated personal radiation dose.

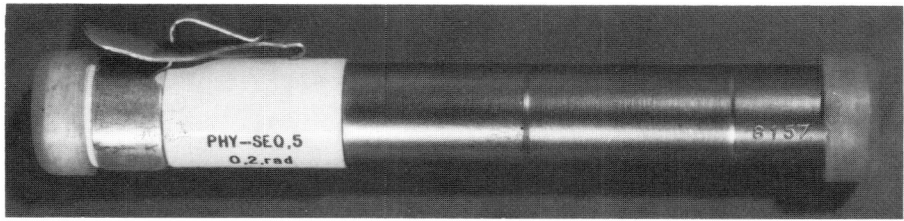

33 Badges for routine monitoring of accumulated personal radiation dose:
A   film badge complete;
B   film insert;
C   thermoluminescent dosemeter;
D   insert wrapped as issued by and returned to laboratory;
E   insert showing two thermoluminescent plastic discs.

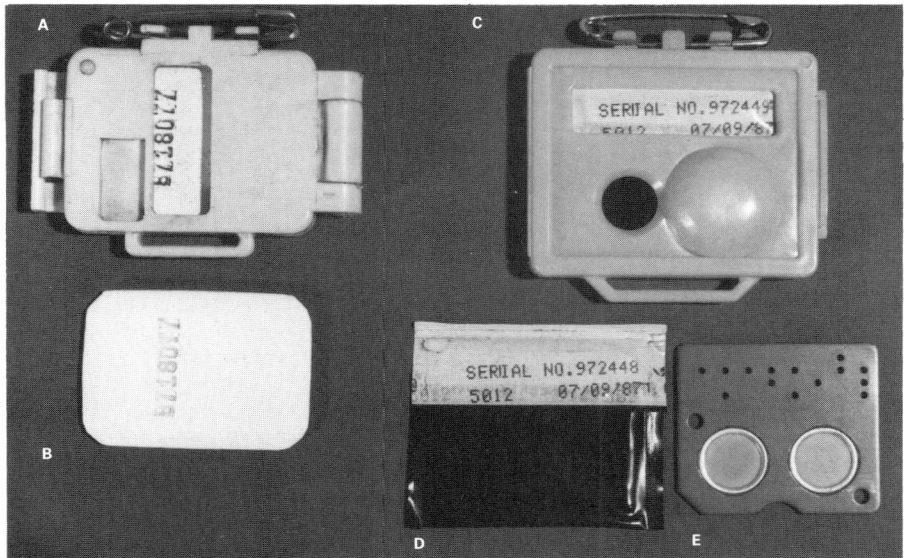

indication of dose. These can be direct-reading, like the quartz fibre electrometer (QFE) (Fig.32), or with predetermined alarm levels for either dose rate or integrated dose. Another type of radiation monitor is the longer term personal monitor, such as the film badge and thermoluminescent badge (Fig.33).

Until recently, classified workers wore film badges from which the films were processed once a month, the film density being a measure of the dose received, but new thermoluminescent badges are replacing the film type. These contain lithium fluoride (LiF) discs which emit light when heated, and the amount of light is a measure of the absorbed dose of ionising radiation. These badges can measure doses both larger and smaller than that measured by a film badge and can be processed automatically.

Other methods of radiological inspection

So far only radiographic film has been considered as the detection/recording material, and it represents by far the most important technique for the following reasons:

(a)   its ability to integrate small doses of radiation over a long period of time, i.e. build up an exposure;

(b)   its image contrast enhancement, because the film gradient is much greater than unity.

Alternative techniques produce the X-ray image on a fluorescent screen, that is a screen which converts spatial X-ray intensities to light intensities.

Nowadays, such screen images are rarely viewed directly. Either the screen is viewed with a very sensitive CCTV camera, or it forms the primary screen of an X-ray image intensifier tube in which, by successive conversions, the image is greatly increased in brightness and again viewed with a TV camera.

The advantages of such a system for weld inspection are:

(a)   the X-ray image is directly available for interpretation, in 'real time';
(b)   no film costs are involved, unless a permanent record is required: the permanent record can be on tape or disc;
(c)   the inspector and his TV monitor can be completely remote from the X-ray apparatus and its radiation hazards;
(d)   a moving specimen can be examined: for example, a rotating specimen.

The disadvantages are:

(a)   the image on the TV monitor is never as sharp and clear as a film radiograph on a viewing screen, and is degraded by the TV line raster. The ability to detect fine image detail is rarely as good as on a film image;
(b)   the equipment is costly: the basic X-ray set is still required and the cost of the image intensifier/TV system has to be added to it;
(c)   the equipment is complex, with potential maintenance problems.

Television-fluoroscopic systems (also called 'real-time radiography') have developed rapidly in recent years through the introduction of computers for image enhancement. The signal from the television camera is digitised by treating the image as a matrix of picture elements (pixels). Several television frames can then be stored, integrated and averaged, to reduce image noise, and further digital enhancement such as contrast control and sharpening is also possible. A greatly improved flaw-sensitivity is therefore obtainable.

Real-time radiographic systems have also been developed which use a microfocus X-ray tube, with which a sharp magnified image of a weld can be obtained, and this procedure can also produce further improvements in image quality by minimising the effects of the image unsharpness. Automated defect recognition, using the computer, is a potential further development.

Other methods of image recording, using powder to display electrostatic charge images, known as ionography, xeroradiography and electroradiography can also be used for radiography but they have not yet found any extensive industrial applications. They have the potential for saving the cost of film and being made into automated processes.

Radiographs can be taken with thermal neutrons, but as far as it is known this has no applications to weld inspection.

For and against radiographic methods

Radiography on film produces a permanent record, with an image which is normally easy to interpret: if there are any problems in interpretation, a second opinion can be obtained. Techniques are well developed for straightforward applications, are well documented, and do not require great skill to apply them. Radiography can be applied to welds in any material — copper, austenitic steels, plastics — up to the thickness limits of the particular equipment, and up to a maximum of about 500mm of steel. However, the radiography of thick ferrous material (> 150mm) requires very expensive equipment, and the laboratories needed to house such equipment adequately are also expensive.

The technical limitations of radiography are its inability to detect, with certainty, tight cracks, small cracks and those lying at an angle to the radiation beam. Moreover, the crack and other planar defects, such as lack of fusion, are potentially the most dangerous defects in welds subject to fatigue conditions and these are the defects which are the most difficult to reveal by radiography, particularly in thick metal sections, where the radiograph is inherently unsharp. If such defects are surface-breaking they can be found by relatively cheap techniques, such as magnetic or penetrant crack detection, and one of these methods should always be used as a supplementary test to radiography in weld inspection.

Supplementary techniques such as stereometry can be used in radiography for defect depth measurement, and microdensitometry of a radiographic image can be used to estimate the through-thickness height of a defect.

References

1   IIW/IIS-335-69: Viewing conditions for radiographs. Publ International Institute of Welding, London, 1969.
2   BS 3971: 1985: Specification of image quality indicators for industrial radiography (including guidance on their use). Publ British Standards Institution, 1985.
3   Halmshaw R: 'Industrial radiography techniques.' Publ Taylor & Francis, London, 1971.
4   ISO 6520: 1982: Classification of imperfections in metallic fusion welds, with explanations. Publ International Standards Organisation, 1982.
5   BS 2600: 1983: Radiographic examination of fusion welded butt joints in steel, 1983.
6   BS 2910: 1986: Methods for the radiographic examination of fusion welded circumferential butt joints in steel pipes, 1986.
7   BS 5650: 1978: Regulations for the safe transport of radioactive materials, 1978.
8   ISO 3999: 1977: Regulations for the safe transport of radioactive materials, 1979.
9   IAEA. 1973: Regulations on public transport of source containers. Publ International Atomic Energy Agency, Vienna, 1973.
10 'ICRP recommendations on radiation units.' *Radiology* 1975 **125** (2) 492.

General background reading

Halmshaw R: 'Industrial radiology: theory and practice.' Applied Science Publishers Ltd, London, 1982.

'Industrial radiography.' Publ Agfa-Gevaert, Mortsel, Belgium, 1986.

'Radiography in modern industry.' 4th edition. Publ Kodak Ltd, Rochester, NY, USA, 1980.

'Handbook of radiographic apparatus and techniques.' 2nd edition. Publ International Institute of Welding, London, 1973.

'Non-destructive testing handbook.' 2nd edition. Vol.2: 'Radiography and radiation testing.' Publ American Society for NDT, Columbus, Ohio, USA, 1985.

3 Ultrasonic methods

Most ultrasonic flaw detection in welds is done by moving a small probe (the transducer) over the surface of the parent plate adjacent to the weld by hand and watching a display on an oscilloscope screen. The probe needs to be 'coupled' to the surface of the metal by a layer of water, oil or grease, and produces a beam of ultrasound which passes into the metal and is reflected back from any weld defect or other discontinuity. This is the 'pulse echo' technique and its success depends upon an accurate knowledge of the beam direction, its size and the physical principles involved.

Basic principles

As the name implies, ultrasonic waves are mechanical vibrations having the same characteristics as sound waves, but having a frequency so high that they cannot be detected by the human ear, i.e. greater than about 20kHz. For weld examination in metals the ultrasonic waves usually have a frequency in the range 500kHz to 10MHz, most applications using a frequency between 2 and 5MHz.

It is important to realise that ultrasonic waves are not electromagnetic radiation passing through the specimen, but are the result of induced particle vibration in the specimen, and are possible because of the elastic properties of the material of the specimen. For this reason, the wave velocity is different in different materials (Table 6). From the basic velocity law:

velocity of waves = wavelength$\times$frequency (V = $\lambda \times$n)

and, taking the velocity of longitudinal waves in steel as approximately 6×10^5 cm/sec, the wavelength of 2MHz ultrasonic waves is 3mm (in steel). There are several types of ultrasonic wave, the most important being compressional (longitudinal), shear (transverse) and surface waves. In compressional waves the particles of the transmitting material move in the direction of wave propagation, and these waves can be transmitted in solids and liquids.

In shear waves the particles of the material vibrate at right angles to the direction of travel of the waves, and shear waves cannot be propagated in liquids. They have a velocity approximately half that of compressional waves in the same material.

Table 6. *Typical ultrasonic wave velocities in different materials, m/sec*

Material	Compressional	Shear	Surface (Rayleigh)
Steel	5.9×10^3	3.2×10^3	3×10^3
Aluminium	6.2×10^3	3.1×10^3	2.8×10^3
Water (20°C)	1.5×10^3	—	—
Perspex	2.7×10^3	1.1×10^3	1.0×10^3
Oil	1.2×10^3	—	—

There are several different types of surface wave. Rayleigh waves can be propagated and are somewhat analogous to water waves, in which the motion of the particles is both transverse and longitudinal in a plane containing the direction of propagation and the normal to the surface. If the specimen thickness is comparable to the wavelength, a type of surface wave known as a Lamb wave can be generated, and as the velocity of propagation is a function of plate thickness and frequency there can be an infinite number of Lamb wave modes. Waves which travel on the surface without any vertical component are known as Love waves. Rayleigh, Lamb and Love waves are used for special applications of ultrasonic testing, but by far the most important types of wave for ultrasonic flaw detection in welds are compressional and shear waves.

Nearly all methods of ultrasonic flaw detection use the pulse echo technique in which a short ultrasonic pulse is propagated from a transmitter probe, through a coupling medium, into the material under test. During its travel this pulse is partially reflected from any discontinuities in its path and the 'echoes' produced are picked up by a receiver probe, which may be the transmitter probe itself (used as a transceiver) or a separate probe. It is quite feasible and common for the transmitter probe to act as a receiver between successive transmission pulses. Any surface where there is a change in elastic properties (cavity, inclusion, segregation) can act as a reflecting surface and give rise to an echo. The echo from the discontinuity provides information about its position and size, the usual method of information presentation being the so-called 'A-scan' on an oscilloscope display screen (see below).

Figure 34 is a schematic illustration of an ultrasonic probe generating a

34 The principle of ultrasonic flaw detection, using a single shear wave probe and A-scan on oscilloscope display screen.

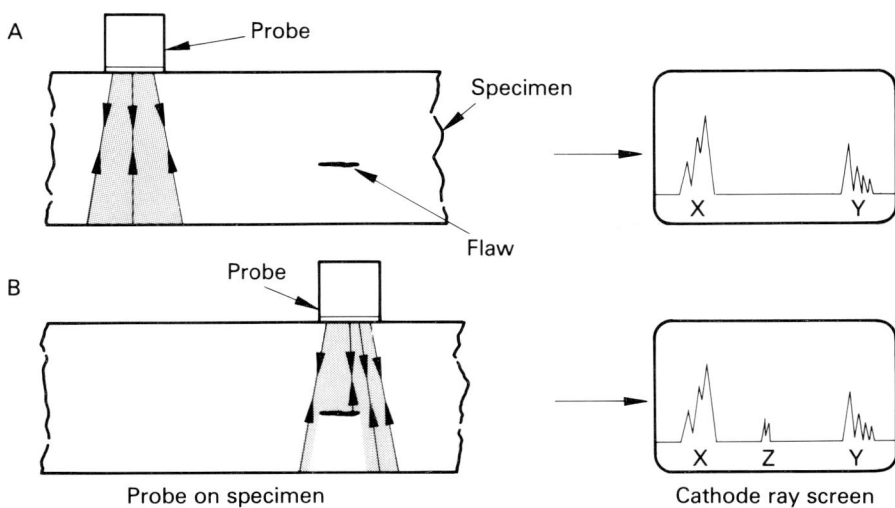

compressional wave. The probe is placed on a flat plate and coupled to the surface by a thin film of grease. When the probe is pulsed, the electronic circuitry produces an indication on the screen at A; the circuit then measures the time the pulse takes to travel through the specimen to the bottom and back and produces a screen indication at B. If some of the ultrasonic beam strikes a cavity, it obviously reflects back to the probe in a shorter time, and a corresponding indication is produced on the screen at C. If the electronic circuitry has a linear time-base, distance AC is a measure of the distance of the cavity from the probe, and the height of the indication can be used as some measure of the size of the cavity.

Piezoelectric crystals are almost universally used as the transducer element in ultrasonic probes. This is a material which, when given an electrical pulse, changes its thickness, i.e. it vibrates and produces ultrasonic waves (the transmission mode). Conversely, when it is caused to vibrate by an incident ultrasonic wave, it produces an electrical pulse (the receiver mode). Piezoelectric materials are used in the form of thin discs with metallised surfaces, the thickness of the disc being related to the natural vibration frequency. In an ultrasonic flaw detection probe the piezoelectric disc is usually damped either mechanically or electrically, as it is desirable to have a very short train of ultrasonic waves in each pulse. Thus, the probe emits a wave packet which has a dominant frequency equal to the natural frequency of the crystal, but with a rapidly decaying train of vibrations.

There are some natural piezoelectric materials such as quartz, but most modern probes use synthetic materials such as lead zirconate titanate (PZT) or lead niobate, which have superior and more reproducible properties. One problem for many years in ultrasonic testing has been the variability of performance of ultrasonic probes of the same nominal construction, and some large inspection authorities have produced detailed acceptance codes for probe performance.[1-2] If ultrasonic flaw detection is to be executed with a high degree of reliability the operator must have a precise knowledge of the performance of the equipment, and the equipment must be accurately calibrated (see section on 'Calibration blocks', below).

Beam spread

Although, because of their short wavelength, ultrasonic waves travel essentially in a straight line, there is always some spread of the beam. The angle of spread, θ, is given by:

$$\sin\frac{\theta}{2} = \frac{1.22\lambda}{D} \qquad\qquad [5]$$

where λ is the ultrasonic wavelength in the material and D is the crystal diameter. Thus, a higher frequency probe produces less beam spread.

An ultrasonic beam from a probe is usually described as having a 'near zone' (or Fresnel region) and a 'far zone' (or Fraunhofer region). The near zone is an approximately parallel-sided beam and its length, N, is given by:

$$N = \frac{D^2}{4\lambda}$$

In this region there are marked variations of maximum and minimum ultrasonic intensities which can cause serious problems in flaw size estimation if the flaw is in this region of the beam. In the far zone the beam diverges and the ultrasonic intensity decreases according to the inverse square law, and it is to this region that Eq.[5] for beam width applies.

Acoustic impedance

When an ultrasonic wave strikes an interface between two media, at right angles, some of the energy is reflected and some transmitted.

If E_i is the incident energy,
$\quad E_t$ the transmitted energy and
$\quad E_r$ the reflected energy,

$$E_t = \frac{4 Z_1 Z_2}{(Z_1 + Z_2)^2} \cdot E_i$$

$$E_r = \left(\frac{Z_1 - Z_2}{Z_1 + Z_2}\right)^2 \cdot E_i \qquad [6]$$

Z_1 and Z_2 are called the *acoustic impedances* of the two materials forming the interface, defined from $Z = \rho \times V$ where ρ is the material density and V is the ultrasonic velocity.

If values are inserted for steel and water, it will be seen that about 12% of the incident energy will pass through a water/steel interface and 88% will be reflected, whereas with air less than 1% is transmitted.

35 *(Below)* Ultrasonic waves incident on interface between two media, 1 and 2, at angle: compressional wave.

36 *(Below right)* Ultrasonic waves incident on interface between two media, 1 and 2, at angle: shear wave.

Reflection and refraction: mode conversion

When an ultrasonic beam strikes an interface between two different materials at any angle but normal it can produce both reflected and refracted compressional and shear waves. The two cases of an incident compressional wave (Fig.35) and an incident shear wave (Fig.36) are shown. Simple relationships describe the angles and velocities of these various waves, the general equation being known as Snell's Law.

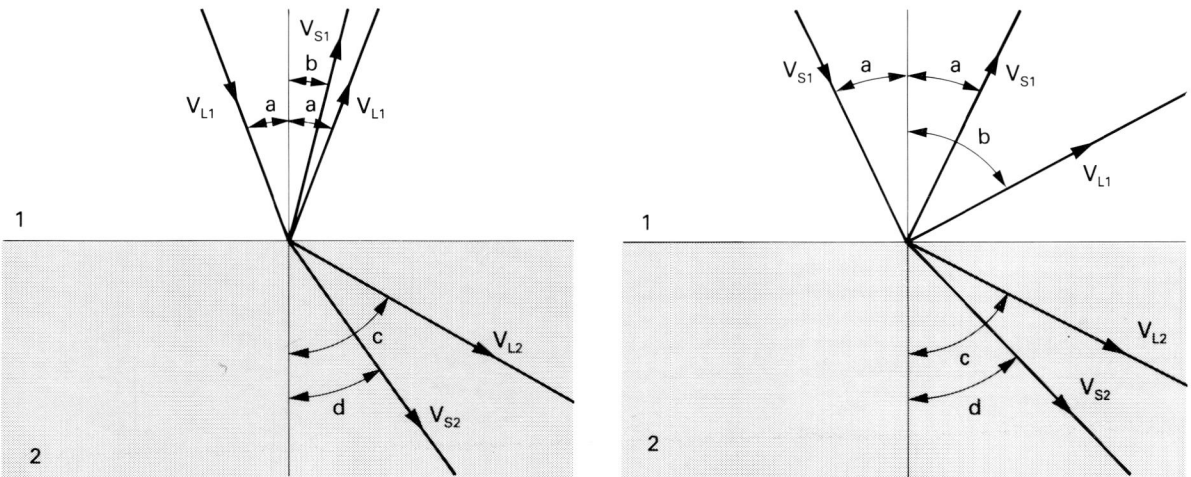

In Fig.35:

$$\frac{\sin a}{V_{L1}} = \frac{\sin b}{V_{S1}} = \frac{\sin c}{V_{L2}} = \frac{\sin d}{V_{S2}}$$

where:
V_{L1} is the longitudinal (compressional) wave velocity in material 1,
V_{L2} is the compressional wave velocity in material 2,
V_{S1} is the shear wave velocity in material 1 and
V_{S2} is the shear wave velocity in material 2.

Exactly the same relationships apply to an incident shear wave (Fig.36):

$$\frac{\sin a}{V_{S1}} = \frac{\sin b}{V_{L1}} = \frac{\sin c}{V_{L2}} = \frac{\sin d}{V_{S2}}$$

and can be applied analogously for a wave travelling from medium 2 to medium 1.

In both cases, depending upon the angle of the incident wave, some of the secondary waves may not exist. Above a critical angle for each incident wave there may be total reflection for that kind of wave, so that for an incident compressional wave there are two critical angles: the first when there is no transmitted compressional wave in medium 2, and the second when neither a compressional nor a shear wave is transmitted.

It is important to remember that this reflection and refraction occurs at any interface, so that in a complex specimen there may be many secondary ultrasonic waves, some of which can return to the receiving probe and produce what are usually called 'ghost' echoes. The relative intensities of the refracted compressional and shear waves vary markedly with angle and are related to the relative wave velocities in the two media. The equations are very complex.[3,4]

An important application of Snell's Law is in the construction of shear wave probes (Fig.37). The disc of piezoelectric material is mounted at an angle on a

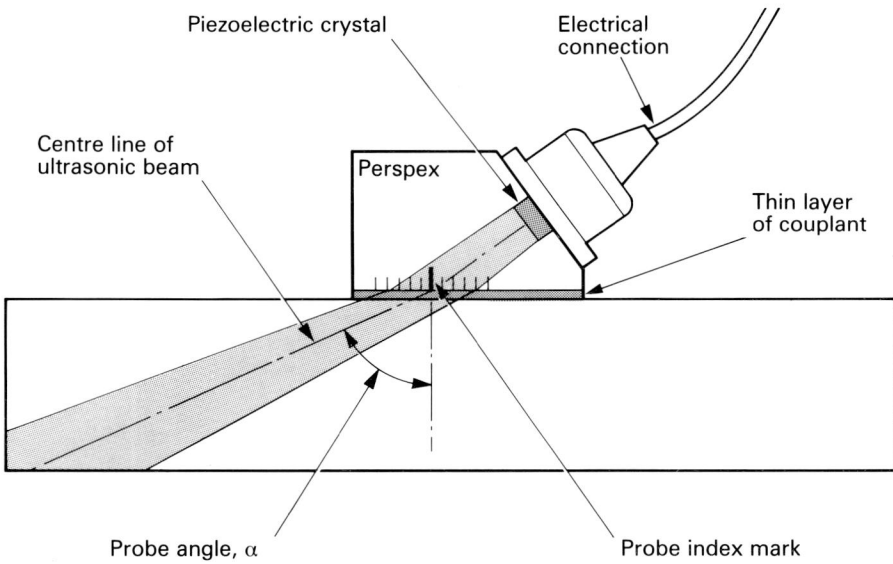

37 Typical shear wave probe showing construction and mode of operation.

Piezoelectric crystal

Electrical connection

Centre line of ultrasonic beam

Perspex

Thin layer of couplant

Probe angle, α

Probe index mark

block of Perspex (the 'shoe'), which in turn is coupled to the surface of the metal specimen with a couplant film. The piezoelectric disc generates pulses of compressional waves by electrical excitation, but the angle of the Perspex is chosen in relation to the relative velocities of ultrasonic waves in the metal and Perspex, so that only shear waves of predetermined angle are transmitted into the specimen.

Probes are commercially available for angles from 35° to 80° for use on steel, this being the angle of the centreline of the ultrasonic beam to the normal in the steel specimen; if the shoe is used on another metal the probe angle will not be the same. The Perspex shoe is often shaped or grooved to minimise the effects of the ultrasonic energy reflected at the Perspex/steel interface and which may otherwise find its way back to the crystal when the probe is used as a transceiver.

Decibel notation

A convenient way to specify variations in ultrasonic intensity arising from attenuation, beam spread, etc. is in terms of decibels, dB. A decibel is one-tenth of a bel, which is a unit of power based on logarithms to base 10, so that if two powers are P_1, P_2, they are said to differ by n bels if $P_1/P_2 = 10^n$; or

$$n = \log_{10} P_1/P_2 \text{ bels}$$
$$= 10 \log_{10} P_1/P_2 \text{ decibels}$$

As acoustic power is proportional to (amplitude)^2, an amplitude ratio of 2:1 corresponds to very nearly 6dB (dB values can be added; amplitudes must be multiplied). Examples:

```
 3dB = 1.41 amplitude ratio
 6dB = 2.00 amplitude ratio
10dB = 3.16 amplitude ratio
20dB = 10.00 amplitude ratio
30dB = 31.6 amplitude ratio
```

Attenuation

When an ultrasonic wave is propagated through material it is attenuated by various mechanisms including scattering. The attenuation depends markedly on the nature and structure of the material (grain size, grain orientation) and is also a function of the ultrasonic frequency. The attenuation of ultrasonic waves in many engineering materials is small, so that waves can travel several metres in fine-grain mild steel or wrought aluminium, but there are difficulties in applying ultrasonic testing to copper-based alloys and to austenitic steel welding. In austenitic steel welds, because of the markedly anisotropic grain structure and the sizes of the grains, ultrasonic beams are heavily scattered, they can bend, and defect echoes appear to originate from false locations (see also 'Ultrasonic inspection of austenitic steel welds', below).

Equipment: manual methods

Successful ultrasonic flaw detection and measurement requires good equipment, which must be very accurately calibrated by the operator. This section will be concerned with the techniques in which the operator moves one or more ultrasonic probes by hand over the surface of a weldment, and the probes are connected by flexible cable to the electronic unit which contains the pulser, amplifier and the display oscilloscope. The display will be basically A-scan, as shown in Fig.34, but using a shear-wave probe as shown in Fig.37. Normally there are no automatic recording facilities and the operator must write down his findings.

An ultrasonic flaw detector (Fig.38) contains a series of units with a block diagram similar to that shown in Fig.39. The time base circuit is initiated by an internal oscillator, which then causes the fluorescent spot on the CRT screen to commence moving from left to right (Fig.40). The pulse generator is then initiated and supplies a sharp electrical impulse to the ultrasonic crystal in the probe, which acts as a pulse transmitter. A pulse of ultrasonic energy is thus sent into the specimen and an initial vertical displacement of the fluorescent spot occurs simultaneously, thus marking the instant at which the ultrasonic pulse commences its travel. While the pulse of ultrasonic waves is travelling through the specimen the fluorescent spot is being moved at constant speed across the screen by the time base circuit. Any of the ultrasonic energy which is reflected back from a surface or a defect is picked up by the probe acting as a receiver. The electrical impulse generated by the crystal is amplified and displayed on the screen at a distance from the transmitted pulse indication which is proportional to the distance of the defect from the probe. The sequence is repeated at a rate known as the pulse repetition frequency producing a continuous image on the CRT screen.

For butt weld inspection, although a compressional wave probe may be used to check on the parent plate in the vicinity of the weld for laminations,

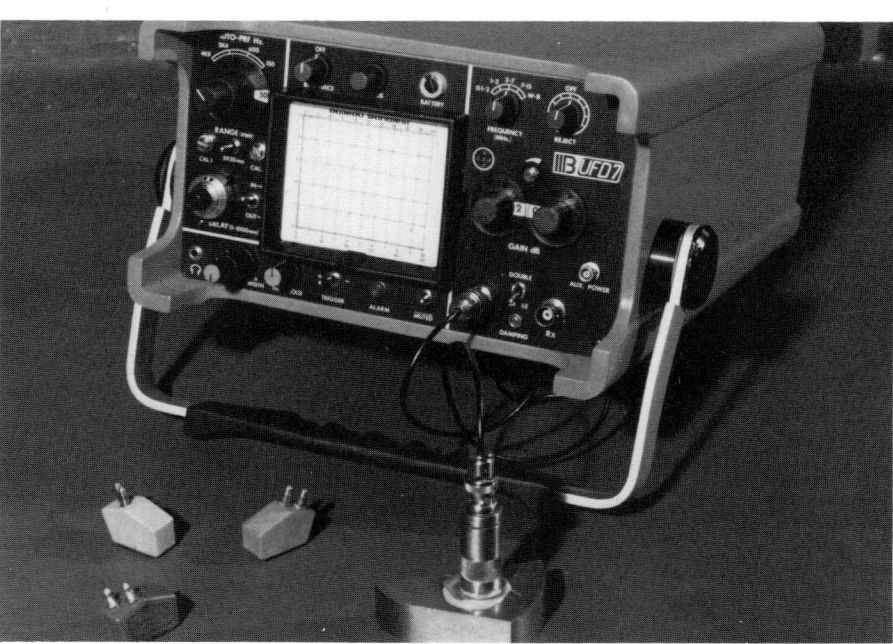

38 Modern ultrasonic flaw detector with a selection of probes.

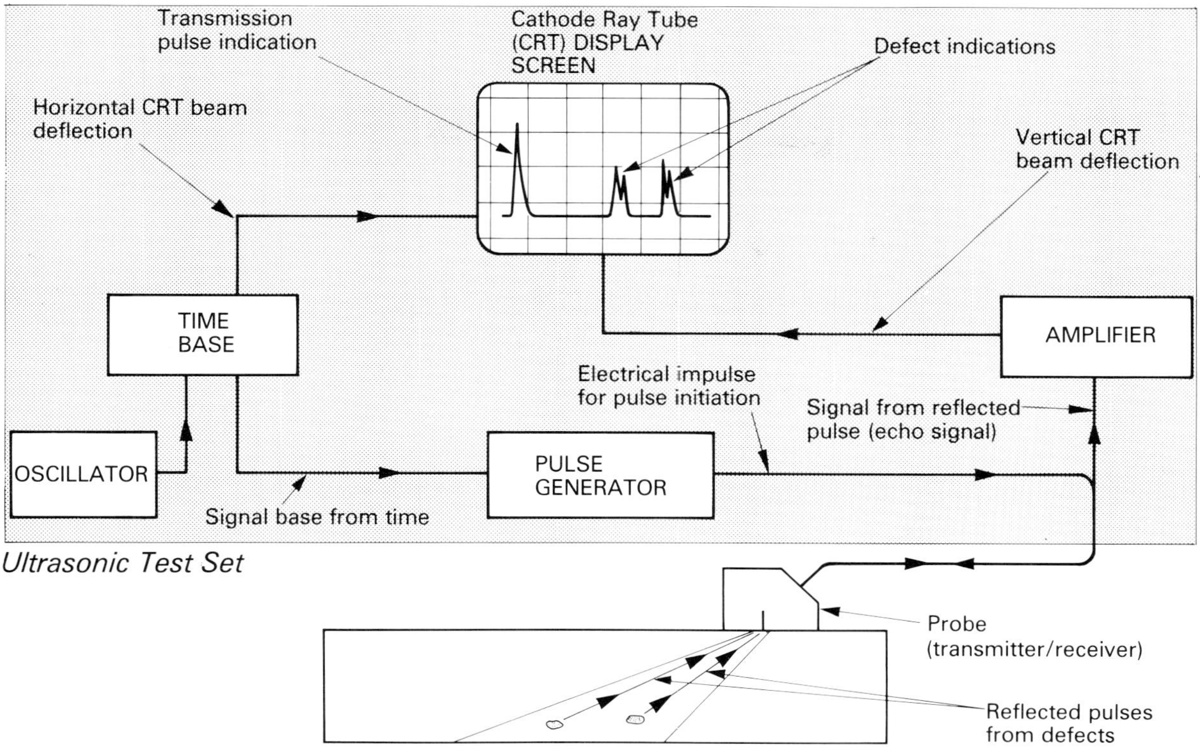

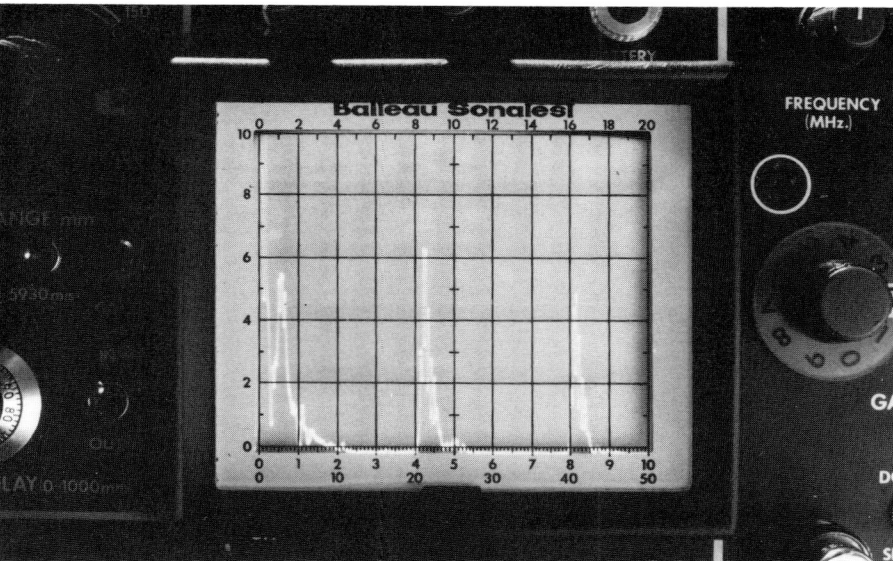

Transmission pulse indication

Cathode Ray Tube (CRT) DISPLAY SCREEN

Defect indications

Horizontal CRT beam deflection

Vertical CRT beam deflection

TIME BASE

AMPLIFIER

Electrical impulse for pulse initiation

Signal from reflected pulse (echo signal)

OSCILLATOR

PULSE GENERATOR

Signal base from time

Ultrasonic Test Set

Probe (transmitter/receiver)

Reflected pulses from defects

39 Block diagram of basic ultrasonic pulse echo equipment.

40 A-scan display on ultrasonic flaw detector screen.

practically all other ultrasonic flaw detection is done with shear wave probes of the type shown in Fig.37.

The minimum requirements for the ultrasonic apparatus are that:

(a)   it is suitable for both single and double probe operation;
(b)   it covers a range of frequencies, at least 1-5MHz;

(c)   the amplifier has a calibrated gain control, in steps of 1 or 2dB;

(d)   suppression of low level signals should be optionally available;

(e)   the time base must be linear within $\pm 1\%$ over its full range;

(f)   the time base range must be adjustable between 100-500mm travel path for both compressional and shear waves;

(g)   it should have a delay or horizontal shift control, so that the zero point can be adjusted with any probe;

(h)   it must have graticules and graduations permanently marked on the display screen, both vertical and horizontal, preferably on the inside so as to avoid parallax effects;

(i)   it must have sufficient trace brilliance for the screen indications to be read without difficulty;

(j)   it should give completely stable indications, after a preliminary warming-up period; if mains operated, there must be built-in compensation for limited voltage fluctuations.

In addition to these basic requirements, most modern equipment has a range of pulse repetition frequencies, scale expansion features, optional electronic distance-amplitude correction, built-in 'gates' of variable width and location for signal extraction, and built-in attenuators for the measurement of pulse height. Some of the equipment characteristics are determined by the probe construction as well as by the apparatus design.

Probes

The simplest type of probe is one which generates or detects compressional waves at normal incidence. This consists of a suitable thickness of piezoelectric material, metallised on both sides and with a backing material bonded to the back; the front surface is usually protected by a thin diaphragm. For weld inspection the probe type shown in Fig.37 for shear waves is the commonest type used. The probe with its connecting cable should not cause any internal noise when operated at the sensitivity required, and all probes should be marked with frequency, probe angle (angle of refraction in steel), crystal material and dimensions, wave type and probe index. It is often necessary to determine additional data on each probe, such as angle of beam spread, near-field length, polar diagram of field, and a series of standard calibration blocks has been devised which enables the characteristics of both probes and equipment to be measured.

Before any work is attempted with new equipment, or after equipment servicing and at periodic intervals during normal use, these additional measurements should be made and recorded. Particularly, probes must be checked before critical work such as defect sizing, as often the correct identification of a weld defect depends on its accurate location to less than a millimetre.

Calibration blocks

There are several widely-used designs of calibration blocks, three of which are illustrated in Fig.41. The A.2 block described in BS 2704: 1983[5] and also known as the IIW No.1 block, is the most widely used. The IIW No.2 block is designed to be smaller and more portable, and is used with miniature probes, while the A.3

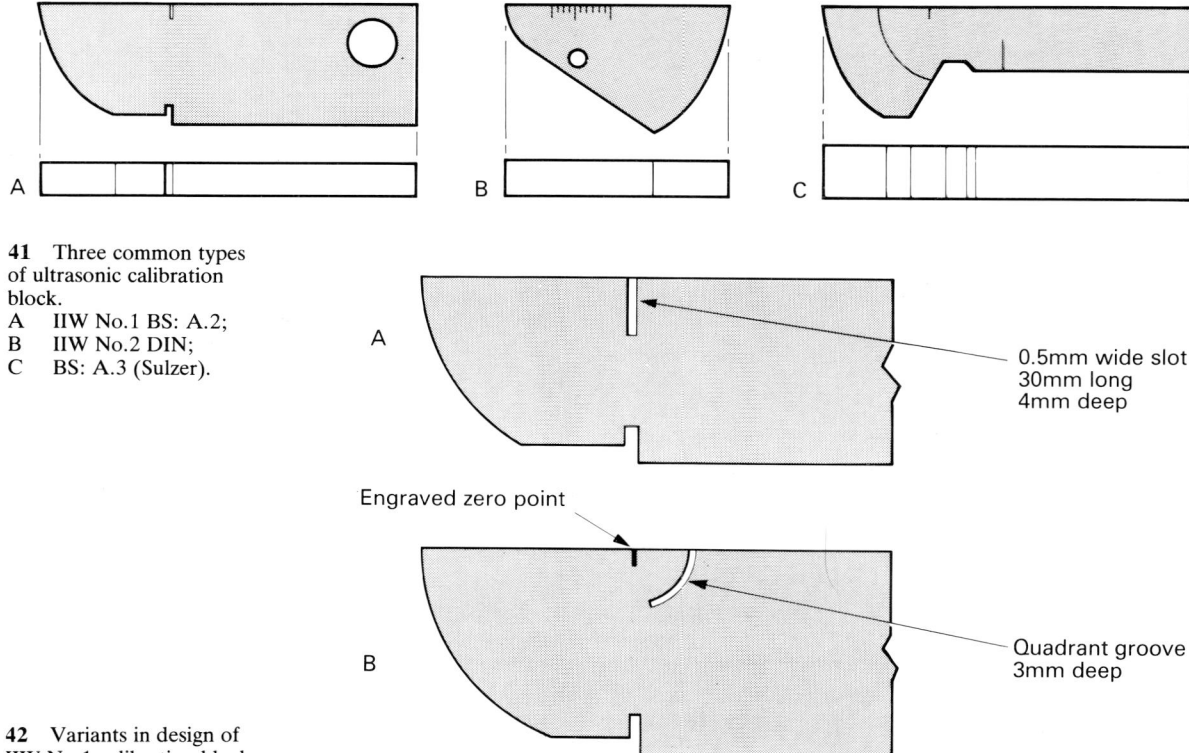

41   Three common types
of ultrasonic calibration
block.
A   IIW No.1 BS: A.2;
B   IIW No.2 DIN;
C   BS: A.3 (Sulzer).

0.5mm wide slot
30mm long
4mm deep

Engraved zero point

Quadrant groove
3mm deep

42   Variants in design of
IIW No.1 calibration block.

block has some extra measurement facilities. For full details of their precise construction, tolerance and surface finish, the relevant Standards should be consulted (BS 2704: 1983;[5] DIN-54-10 1973;[6] ISO-2400-1972;[7] ISO-7963-1985[8]). The A.2 block, which has three versions (Fig.41 and 42), has as its distinguishing feature an additional slot used to provide a reflector for time base range calibration with shear wave probes. The particular advantages of each pattern of calibration block, as well as the detailed procedures for their use, are given in the TWI handbook, *'Procedures and recommendations for the ultrasonic testing of butt-welds'* (1971)[9] and also in the International Institute of Welding handbook, *'Ultrasonic examination of welds'* (1977).[10]

In addition to these standard calibration blocks, some organisations have designed blocks for special requirements, particularly for use in ultrasonic testing of specimens for very large size.

The methods of making the various tests on the equipment and probes are detailed in the above two books and in a range of Standards,[5-8,11] and these will not be described in detail. Tests are desirable on:

· linearity of time base;
· calibration of time base;
· pulse length;
· probe angle;
· probe index (the geometrical centreline of the ultrasonic beam as it emerges from the shoe of the probe);
· beam profile (i.e. the radiation pattern across the beam);
· resolution (in depth);

· correction of transmission point (i.e. the instant in time when the pulse enters the specimen);
· sensitivity setting.

The calibration measurements are particularly necessary because probe shoes can become worn with use, and this wear may not be even, so both probe index and probe angle may be affected. Some organisations, notably the Electrical Supply Industry (ESI), are now specifying accuracies required on these measured parameters. For example, for probe angles (beam angle) of less than 45°, the accuracy must be better than ± 1.5° and the index point within 1mm. Some probes have a squint; that is, the angle between the side edge of the probe and the projection of the beam axis on to the plane of the probe face is not zero. For single crystal probes the squint must not exceed 1.5°.

The signal-to-electronic noise ratio can be measured and should be 70dB weaker than the echo from the 100mm radius of the A.2 block. The presence of side lobes on the main ultrasonic beam and internal echoes in the probe can also be checked.

The approach to flaw location in welds

Before commencing any ultrasonic inspection of the weld it is usual to make a preliminary examination and also to obtain some basic data on the weld. It is necessary to know the welding process and procedure used, the material and its heat treatment, and in particular the joint preparation, especially at the root, and to know whether any repairs have been made. This information should be checked against the weld itself, and this is an appropriate point to locate and mark the centreline of the weld. Accurate knowledge of this line is essential to good inspection.

The normal inspection method is to scan the weld with a probe on the parent plate at the side of the weld, and this area must be free from weld spatter, loose scale or paint, ridges and grooves. A surface roughness better than 6μm RMS is desirable in the area. It is also necessary to check the parent plate ultrasonically in this area, close to the weld, for laminations or inclusions or abnormal attenuation which might give erroneous indications or prevent subsequent normal inspection. The thickness of the parent plate can also be checked at this stage. Normal compression waves are usually used for this parent metal inspection, with either single- or twin-crystal probes.

Surface finish

The weld surface preparation may come into one of four categories (Fig.43):

(a)   as-welded;
(b)   partly dressed to a smooth contour;
(c)   partly dressed to a near flat surface finish;
(d)   fully dressed.

In the as-welded condition, ultrasonic echoes from the weld cap may prevent detection of defects in or near the weld surface. The ideal situation for inspection is (d), but this may be too costly to produce.

43 Surface condition of butt weld:
A   as-welded;
B   partly dressed to smooth contour;
C   near flat surface finish;
D   fully dressed.

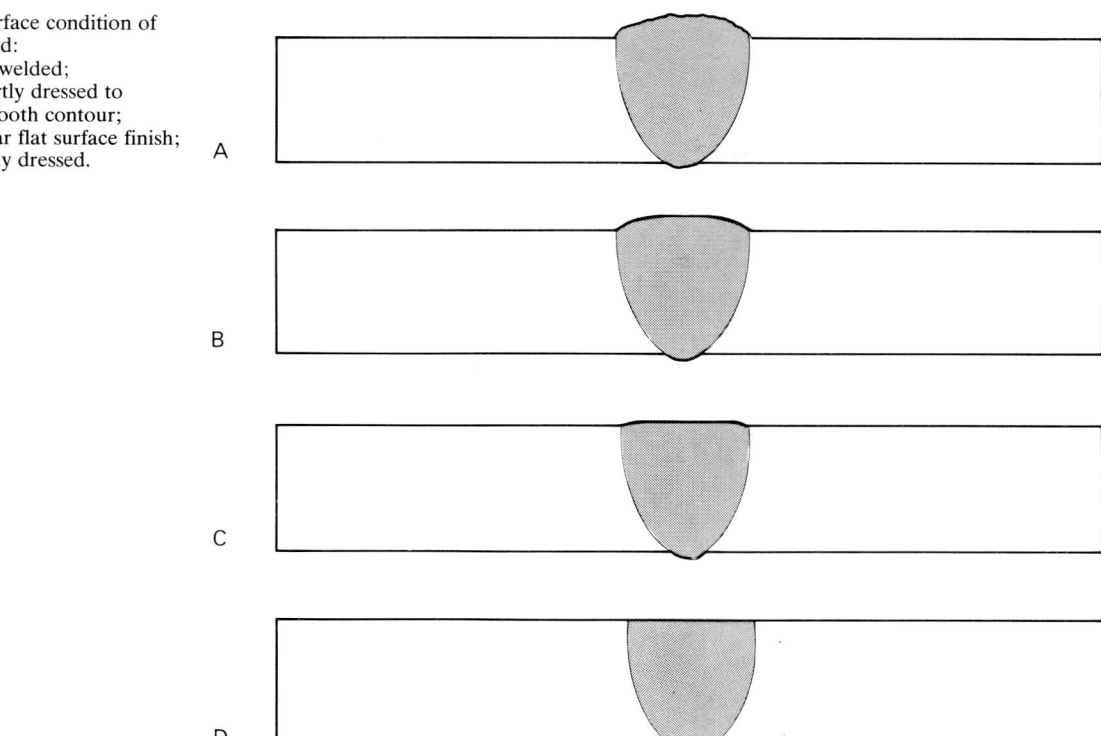

A

B

C

D

Couplants

A couplant, usually a liquid or semi-liquid, is required between the face of the probe and the surface of the specimen to permit transmission of the ultrasonic energy into the specimen. Typical couplants are water, oil, grease, and glycerine; when glycerine is used a small amount of wetting agent should be added. The couplant used should form a film between the probe and the test surface. The viscosity of the couplant used may vary with the surface finish and whether the surface is vertical or horizontal.

Selection of probes

Probe selection is a matter of compromise between many factors: a low frequency is better on a rough surface from the point of view of penetration combined with flaw sensitivity; low frequency waves do not scatter as readily as those of high frequency, but generally the frequency should be as high as possible consistent with adequate transmission, so as to get a better flaw sensitivity and resolution; large probes of low frequency allow more rapid scanning over an area.

Probe angle is selected to ensure that an echo signal will be obtained from all significant flaws — particularly planar flaws, such as lack of fusion on side-walls and at the root — and cracks. To obtain a good echo the angle should be chosen so that the ultrasonic beam will impinge on such flaws as near to normal to the plane of the flaw as possible (Fig.44). Case B will produce a stronger echo signal than case A, which may not produce any signal at all as the return pulse from the flaw will miss the probe. The probe angles most commonly employed are 45°, 60° and 70°, a 70° shear wave probe being suitable for most purposes within the thickness

44 Effect of ultrasonic
beam angle in relation to
detection of planar flaw.
Reflected beam of probe A
will not return to probe.
Probe B, placed further
away, receives the reflected
beam along the same path
as the transmitted beam.

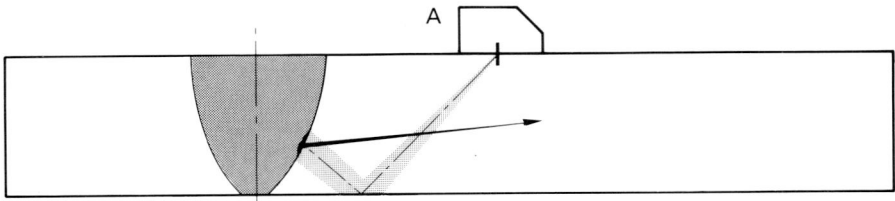

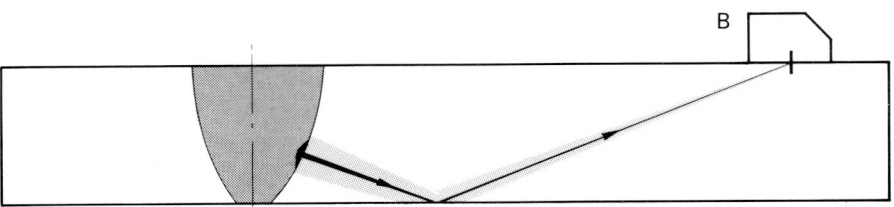

range 10—35mm. For weld thicknesses over 35mm both 45° and 60° probes are used. On curved surfaces probe angles from 35° to 80° may be necessary. A few manufacturers make variable angle probes, but to date these have not been widely used.

Probe scanning

To detect all possible defects the weld metal and the heat-affected zone (HAZ) must be examined over the whole cross-section and along the full length of the weld. Having determined the weld centreline it is therefore necessary to mark off the full-skip and half-skip distance on both sides of the weld (Fig.45). The skip distance is the distance measured on the surface of the weld between the probe index point and the point where the beam reaches the surface after a single reflection from the opposite surface. As will be seen in the figure, if the probe is at full-skip distance the centreline of the ultrasonic beam is reflected from the lower surface of the plate to reach the upper part of the weld, and any defects in this region will reflect ultrasonic energy back along the same path. At half-skip distance, the beam goes directly to the root (bottom) of the weld, and at intermediate distances the ultrasonic beam goes to intermediate regions in the body of the weld. Because of the width of the ultrasonic beam, the whole weld thickness can be covered.

 Scanning may be carried out at up to half-skip distance, or between half- and full-skip distances. If the latter method is used, the test range should not exceed 200mm for 4—6MHz probes or 400mm for 2—3MHz probes. Testing between half- and full-skip distances is the more usual method, but for defect sizing it is desirable to work at a range not exceeding half-skip distance if the thickness and surface finish permit. The full-skip distance is given by (2t×tanα) and the half-skip distance by (t×tanα), where t is the plate thickness and α is the probe angle. It should be noted that with large angle probes and thicker plates the full-skip distance becomes quite large, and there may be both access and attenuation problems.

 To scan the full weld cross-section the probe must be moved repeatedly

45 Stand-off distance
from weld centreline for:
A   half-skip;
B   full-skip;
C   one-and-a-half skip.

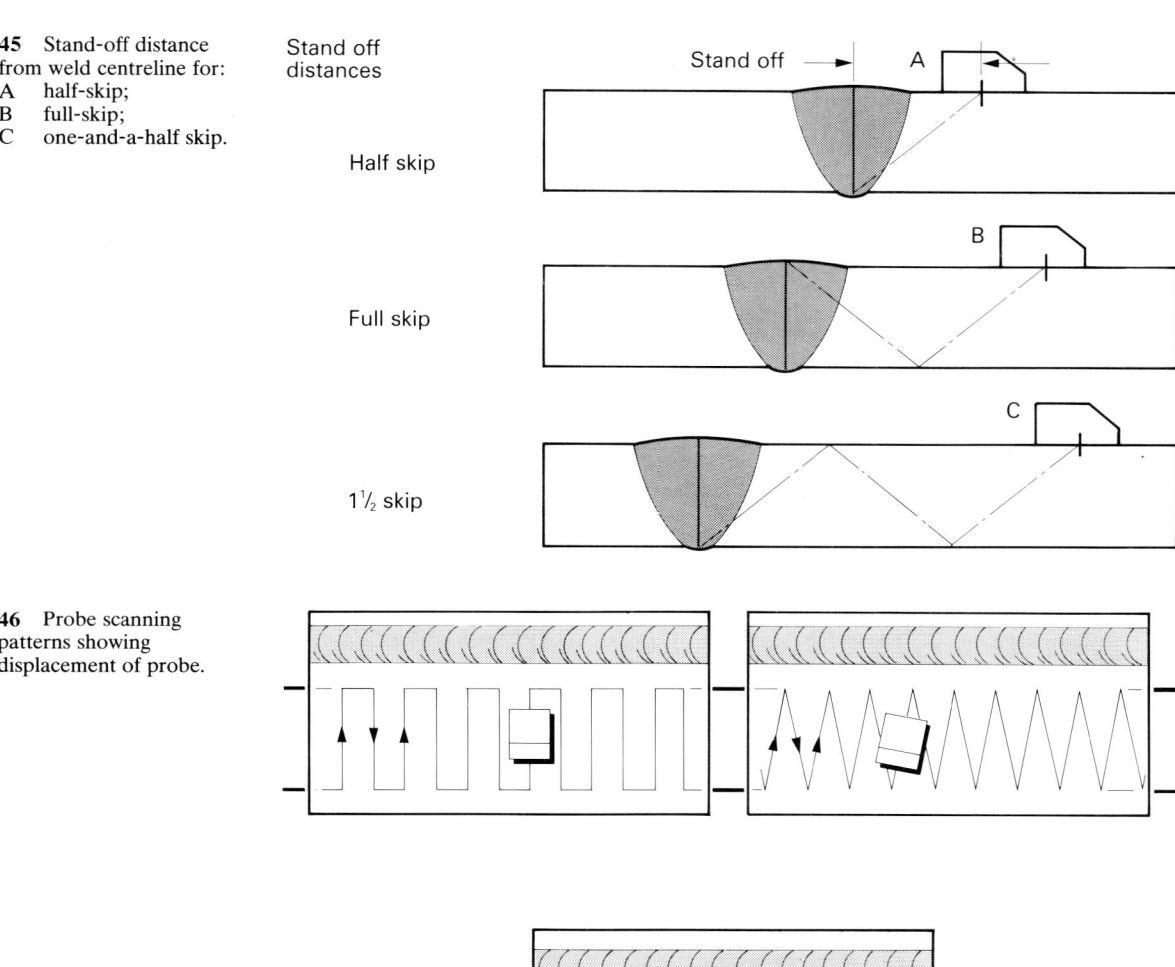

Stand off
distances

Half skip

Full skip

1½ skip

46 Probe scanning
patterns showing
displacement of probe.

Half skip —

Full skip —

from A to B (Fig.45), and during this movement it should be rotated slightly, say
± 15° to the normal. The probe is then worked along the length of the weld. This
scan should be done from both sides of the weld. Typical probe scanning patterns
are shown in Fig.46. If the square scan is used, the displacement of the probe
along the weld should be less than the width of the crystal probe. Welds above
about 150mm thick may need to be scanned from both sides of the weld (Fig.47,
A and B) and both surfaces (B and D) using the single traverse technique, i.e.
with the half-skip range.

     If the weld is accessible from only one side and one surface, a triple traverse
technique must be used (Fig.48). In triple traverse scanning, when the probe is
farthest from the weld the ultrasonic beam is reflected successively from two
surfaces of the welded plate. This method may also be necessary on thin plates
because the probe may meet the weld reinforcement before it can reach point A
on Fig.45.

47 Thick weld inspection: it is necessary to scan from both sides of the weld (A and B, C and D) and both surfaces of the plate (A, B and C, D).

48 The use of double and triple traverse scanning.

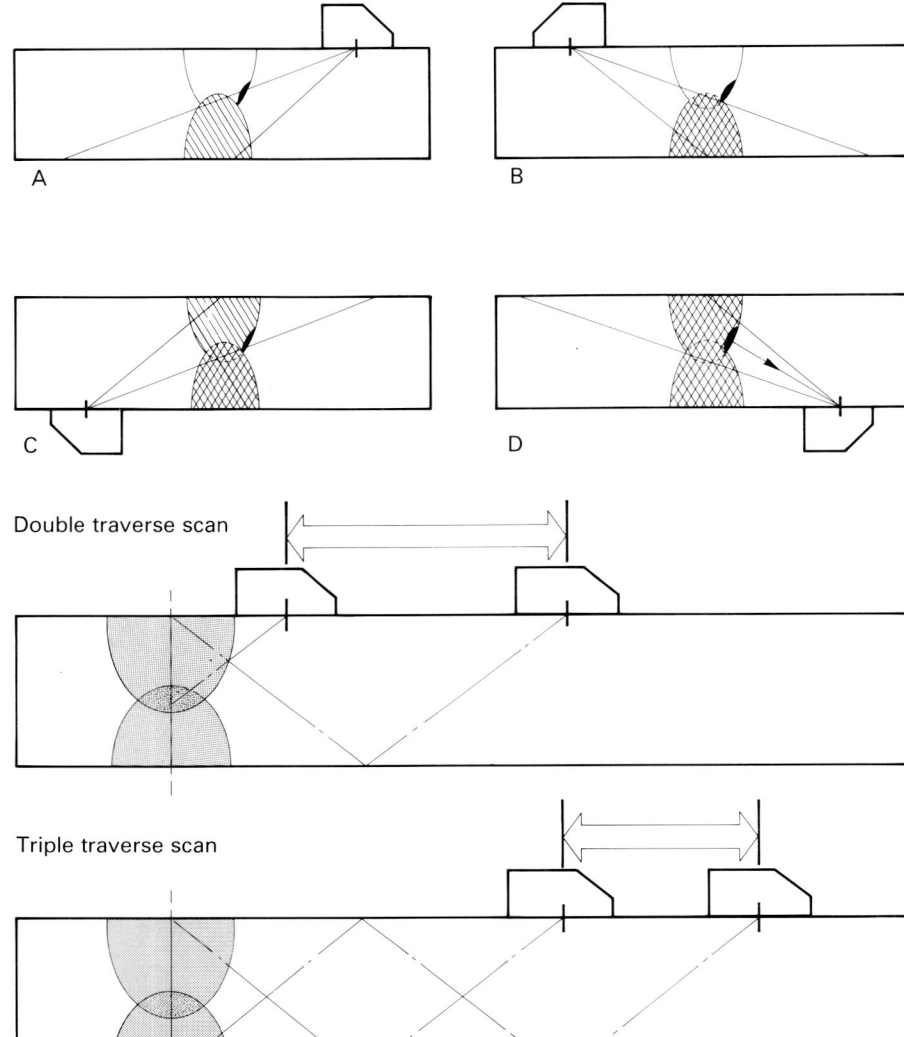

Supplementary scanning. Flaws transverse to the weld direction may not be detected during normal scanning from the side of the weld, and to reveal such defects as transverse cracks the ultrasonic beam must also be directed along the length of the weld. If the weld has been fully dressed, a shear wave probe can be placed directly on the weld surface and moved along in both directions on the centreline. For thick welds a tandem technique is recommended (see next section).

Welds having a penetration bead or backing strip require particular attention to be paid to the root region, and the presence of the bead or backing strip can give rise to echoes which may mask those from root cracks or incomplete root fusion. Consequently, the need for very precise calibration, and prior calculation of range and probe stand-off distance, applies with added force to the root scan on these types of weld.

Probe angles of between 45° and 60° are the most suitable, but larger angles may be required on thin welds because of the short beam path, as the root region must lie outside the near field of the probe.

It is essential to know the exact position of the root faces, before testing, so that the source of an echo signal from the small but critical root zone may be precisely located. A reference datum should be marked on each side of the joint before welding at a known distance from the root face of the joint. It is desirable to move the probe along a guide strip, as shown in Fig.49, having first calculated distances R and S (i.e. knowing the beam angle and probe index point of the probe being used). Even with very careful calibration it is often difficult to distinguish between flaws on the far side of the root and echoes from the penetration bead. An echo derived from root geometry is likely, however, to be more nearly continuous than a flaw signal as the probe is traversed along the weld.

Tandem probe techniques

A number of techniques, using two probes instead of one, are sometimes necessary for welds which are likely to contain flaws perpendicular to the weld surface. Figure 50 shows this method (transmitter, T, receiver R) as applied to

49   Examination of weld root by half-skip distance scanning with a guide strip: R — range from probe; S — stand-off distance from weld centreline.

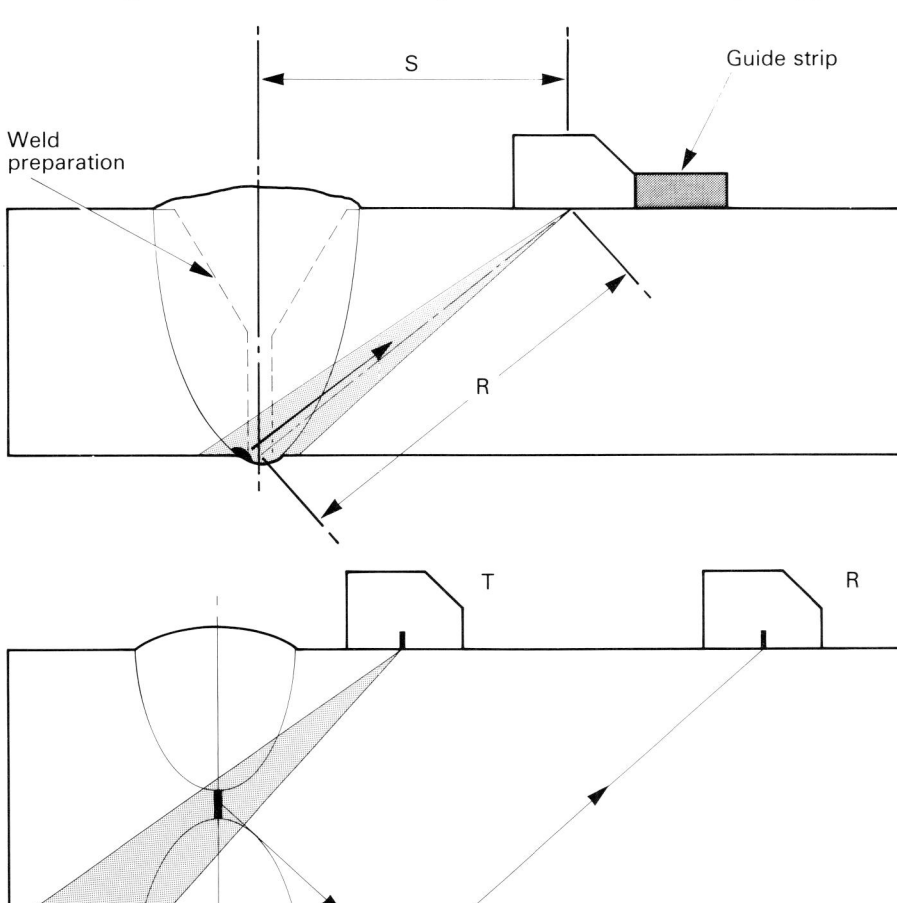

50   Tandem probe technique for the detection of lack of root fusion in a double V weld: T — transmitting probe; R — receiving probe.

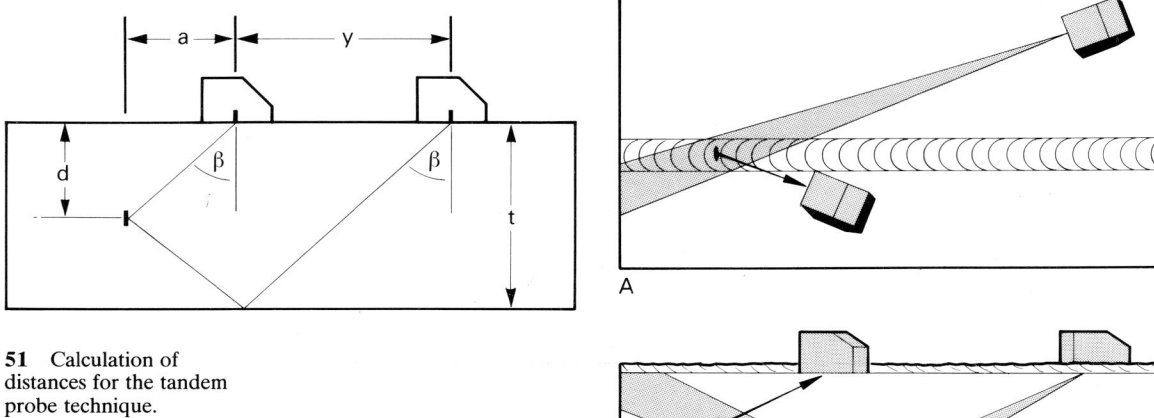

51 Calculation of distances for the tandem probe technique.

52 Examination of transverse defects in a thick weld using tandem technique with the probes straddling the weld bead:
A   plan view;
B   side view.

the detection of lack of root fusion in a double V weld. To avoid having the two probes very far apart, a probe angle between 35° and 55° is usually used; although there are theoretical advantages in sometimes using two probes of different angles, particularly on thick welds, it is more usual to use two closely matched probes. The physical problems in manipulating the two probes in a controlled manner are considerable: usually they are mechanically linked, with a loose linkage to allow a slight rotational movement.

For the tandem technique the time base range is calibrated in the same way as for a single probe and simple formulae can be derived for probes on a flat plate (Fig.51):

$$a = t \cdot \tan\beta - \frac{Y}{2}$$

$$d = t - \frac{Y}{2} \tan\beta$$

Tandem probes so far have been assumed to be in the same plane but when the probe cannot be used on the weld surface, a straddle technique is used with the probes on either side of the weld bead or obstruction (Fig.52). A variation of this technique, popularly known as 'pitch and catch', is to have the two probes a fixed distance apart so that the ultrasonic signal between them, after one bounce on the opposite side, is approximately maximised. As the probes are moved along the weld any flaw entering the ultrasonic beam reduces the energy passing between them. Thus, a defect size indication is obtained which is not dependent on flaw reflectivity.

Sensitivity

As the amplifier of an ultrasonic flaw detector is turned up, the pulse height from a chosen reflector increases in height on the display screen, until eventually the background noise ('grass') also begins to increase, and if the process were continued the grass would swamp all signal indications. An appropriate operating sensitivity level setting must therefore be decided upon. In principle, this setting should be sufficient to show the smallest flaws which it is required to find when these are located at the maximum test range, irrespective of material properties, surface condition or flaw orientation. In normal flaw detection the testing sensitivity used is the setting at which grass from the weld zone, at the appropriate range on the detector screen, is just apparent — say 2mm high average — and for reproducible and quantifiable results this is related to a reference sensitivity which can readily be reset using a standard calibration block.

Narrow cylindrical holes drilled in a block perpendicular to the ultrasonic beam direction and parallel to the scanning surface, form a convenient and cheap-to-construct sensitivity-setting device. The side of the hole is used as the ultrasonic target. In some specifications for ultrasonic testing, the test sensitivity is laid down, for example, in terms of a dB attenuator setting to give a full-screen height echo from the 1.5mm drill hole in the calibration block, or from a series of 3mm diameter side-drilled holes. The sensitivity setting should be referenced to an amplitude correction curve to allow for the variation of flaw-sensitivity with flaw distance from the probe (see, for example, BS 3923: Part 1: 1986).[12]

Several reference reflectors have been used to quantify the sensitivity setting, and some of these are also suitable for flaw size estimation procedures (see below). With angle probes, the 100mm quadrant of the A.2 calibration block (IIW No.1 block, Fig.41a) or the 50mm quadrant of the DIN block (IIW No.2, Fig.41b) can be used, putting the probe at the centre of the curvature; this surface can then be regarded as an infinitely extended flat reflector. For the tandem technique the back surface of a flat, parallel-sided block of appropriate thickness can be used.

The use of drill holes as sensitivity setting devices was referred to above. The 1.5mm hole in the A.2 block is suitable, although a wider block to prevent interference by sidewall reflections is to be preferred, and generally a series of 3mm diameter holes at different distances is used. The end faces of flat-bottomed holes are still used in some industries and some countries, but have caused many problems because of the difficulties in fabrication: very slight variation in the flatness of the hole bottom can cause considerable differences in ultrasonic echo height from the same nominal hole size.

Notches have been used as reference targets but, besides geometric variations, the ultrasonic echo height depends upon the angle of incidence, and notches are not recommended as reference targets.

Finally, there is a method of defect sizing, known as the DGS system, which requires a detailed method of sensitivity setting; this is described in more detail below under 'Flaw size estimation'.

Flaw location and identification

It is a cardinal principle of flaw location that the tester must be able at all times to visualise accurately the path followed by the ultrasonic beam and the beam width; this is the purpose of the calibration procedure described. For fuller details, the reader is referred to the procedural handbooks[9,10] and to the British and other national Standards. To assist this beam visualisation, either diagrams of weld cross-sections must be employed or a flaw location slide with appropriate scales and cursors can be used (Fig.53 and 54). The outline of the weld is drawn in wax pencil on the cursor, together with the beam axis and the 20dB isobars on the base plate. When a flaw echo is received the probe is moved backwards and forwards·at right angles to the line of the weld until the maximum response is obtained, and the distance of the probe index from the centreline is noted and

53 Flaw location slide
(SANDT design, as
supplied by The Welding
Institute):
A inner beam plotting
   card;
B outer transparent
   envelope.

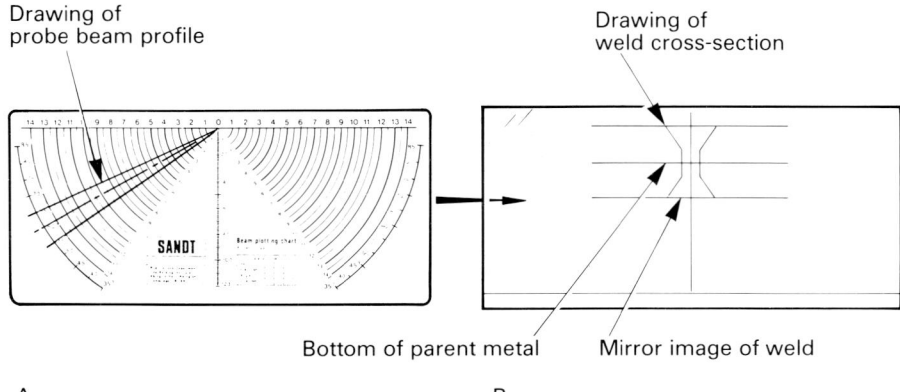

Drawing of
probe beam profile

Drawing of
weld cross-section

Bottom of parent metal   Mirror image of weld

A   B

54 Using the flaw location
slide:
A locating direct echo;
B locating an echo
   reflected in lower
   surface of parent metal;
C actual path of beam
   shown in B.

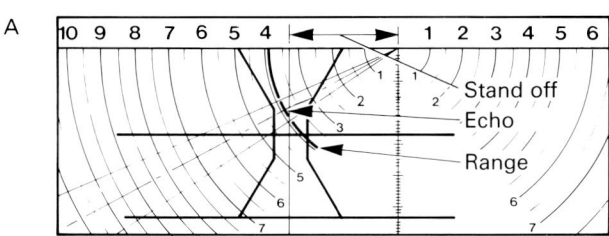

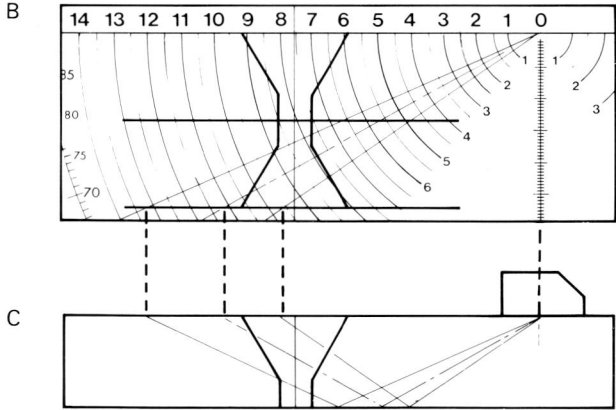

transferred to the slide. The centre of the flaw can then be determined. Fuller
details of the procedure are given elsewhere.[7-10] Although the procedure takes
time, it eliminates guesswork and leads to greater certainty.

   If, when a flaw has been accurately located, the probe is moved slightly both
transversely and parallel to the weld, and rotated slightly, the change in the flaw
pulse echo on the screen can often provide considerable help in identifying the
flaw; such manipulations should be performed from both sides of the weld. The
exact location of the flaw often provides additional information on its likely
nature.

   The echo signals obtained from planar flaws such as cracks and lack of
sidewall fusion are very directional: orbiting the probe from the position of
maximum echo height will result in a sudden drop in pulse height. Different angle
probes also produce large differences in echo height. Slag inclusions, because

they have ragged surfaces, show a small envelope of echo signals, and an orbital probe movement causes little change in the envelope. Gas pores and cavities produce low amplitude signals which remain almost constant as the probe is moved.

Flaw size estimation

The measurement of flaw size in ultrasonic testing is one of the most important, not yet completely solved, problems. Several techniques can provide some information on flaw sizes, but most of these are still being developed and refined. Only in very favourable circumstances, and with additional knowledge of flaw type and orientation, can the size of the ultrasonic echo be taken as a simple measure of flaw size, because of the effects of flaw shape, orientation, attenuation, coupling, distance, resolution, etc.

If the ultrasonic beam can be scanned across a flaw, and if the flaw is large enough, the flaw boundaries can be determined by the distance the probe must move each side of the maximum echo position for the echo to fall to a predetermined value. Several methods have been proposed. In the 6dB drop method the assumption is that, if the axis of the beam is moved so that it is over the edge of the flaw, the response will fall by one half (i.e. 6dB); thus, the two edges of the flaw can be found. This method works reasonably well for very large flaws — larger than about twice the beam diameter — but less well for smaller flaws because of the relatively large beam width from a typical ultrasonic probe.

The 20dB drop method is similar in basic principles, the argument being that, when the maximum response has dropped by 20dB, the flaw is virtually outside the ultrasonic beam; thus, the distance of movement of the probe, from one side of the flaw to the other, is the flaw size plus the beam diameter. Again, errors arise because the ultrasonic beam does not have sharp cut-off edges, but good accuracy on flaws as small as 5mm diameter has been claimed.

A third method of flaw size estimation is by comparison with known reflectors in test pieces or calibration blocks. To be realistic, the test pieces must match the specimen in size, attenuation, surface finish and coupling conditions. The fourth method, and perhaps the most widely used, is the evaluation of defect echo amplitudes by comparison with DGS (distance-gain-size) scales, sometimes known as Krautkrämer AVG diagrams (Fig.55). These diagrams,

55 Example of AVG (DGS) diagram:
A distance, near zone units;
V amplification, dB;
G ratio of flat-bottomed hole diameter/crystal diameter.

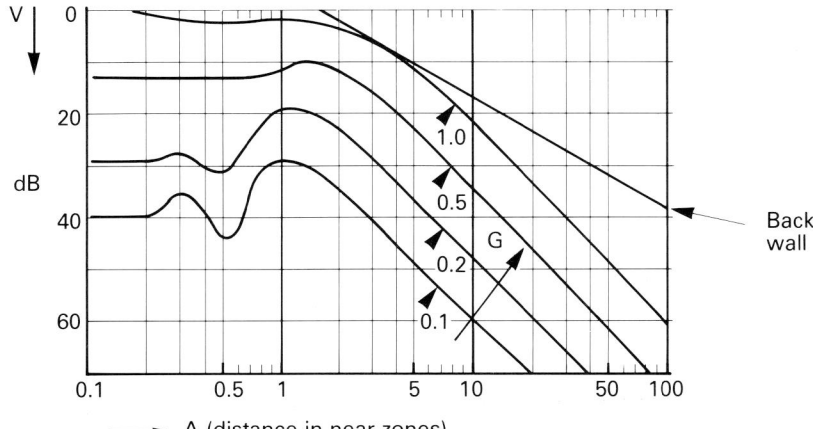

which must be developed for each equipment and probe, express the flaw echo in terms of an equivalent value for a circular flat reference reflector at normal incidence, at the same distance as the flaw from the probe. The value obtained is not therefore true flaw size, but the size of a perfect reflector — i.e. the flaw cannot be *smaller* than this size. DGS diagrams for specific probes can be developed from a general diagram and checked experimentally with a measured back-wall echo, or the echo from a standard reference flaw. Such diagrams may need correction for attenuation losses and also for differences in coupling conditions between the calibration block and the specimen. The system can be much simplified by the use of scales mounted in front of the display screen for particular probes.[10] These DGS diagrams can be developed for tandem probe systems, and with some ultrasonic equipments having built-in distance-amplitude correction they can be very simple. Important new methods of flaw sizing such as time-of-flight diffraction (TOFD) are described in the section on 'Special techniques', below.

Reporting

It should always be remembered that the level of pulse height required for reporting purposes, as given in various specifications, does not necessarily require *all* flaw echoes to be reported, nor is it necessarily the level for acceptability of defects. The acceptability level must also take account of the flaw length, its nature, and its location. The method of reporting a flaw pulse height is normally in terms of the attenuator reading to bring the pulse height to a predetermined level. The report must contain all the information on probe position, equipment characteristics and geometry, to enable an examination to be repeated, and coding systems and standard formats have been proposed to simplify reporting. The previous sections have been written in terms of the manual ultrasonic testing of butt welds in flat plates, but all the methods described can be and have been modified to be applied to curved plates, pipe welds and fillet welds, using exactly the same principles (see next section).

These techniques can be applied to most ferritic materials except austenitic stainless steel welds, which are a separate and difficult problem (see below) and to most light alloy weldments. At present, welds in copper-based metals also present difficult problems in ultrasonic testing because of anisotropy in the grain structure, and the inspection of electroslag welds can be done successfully only under limited conditions.

The essentials of the technique are the very accurate setting-up and calibration of the equipment so that the operator knows exactly the size and direction of the ultrasonic beam which he is using; the use of slide rules or diagrams based on accurate knowledge of the weld configuration and location, to eliminate guesswork; experience and training, to interpret the changes in flaw echo with probe movement; a thorough knowledge of the welding procedure, likely defects and their location. Perhaps the most important requirement is a full-size accurate cross-section of the weld as it exists, on which the ultrasonic beam can be drawn, and this is particularly important in such applications as nozzle welds in which the cross section changes as the probe moves around the weld.

Applications to other than butt welds

On fillet welds ultrasonic testing is mostly used to check the extent of penetration, or the gap between welded components, on a partial penetration weld. Figure 56 shows a typical application in which the probe is kept at a fixed distance from the weld, preferably using a probe guide bar; the optimum distance is full-skip, but needs to be established empirically. If the same joint is designed as a full penetration weld (Fig.57) it may be necessary to make a tandem scan with shear wave probes (1 and 2) as well as a scan with a compressional wave probe (3). Generally, 70° shear wave probes are suitable but to detect HAZ cracking 45° probes would be used (Fig.58).

56 Examination of partly penetrated T-weld, using guide strip to maintain a constant distance, D, from the probe to the root of the weld.

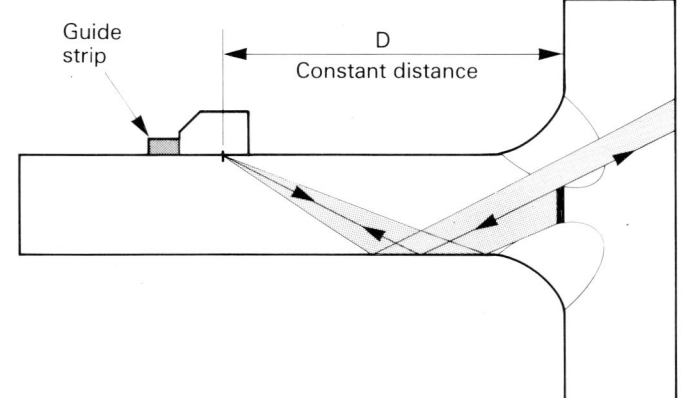

57 Examination of T-weld for lack of root penetration. Probes at positions 1 and 2 are used as a transmit/receive pair. Probe at position 3 is a normal compressional wave probe.

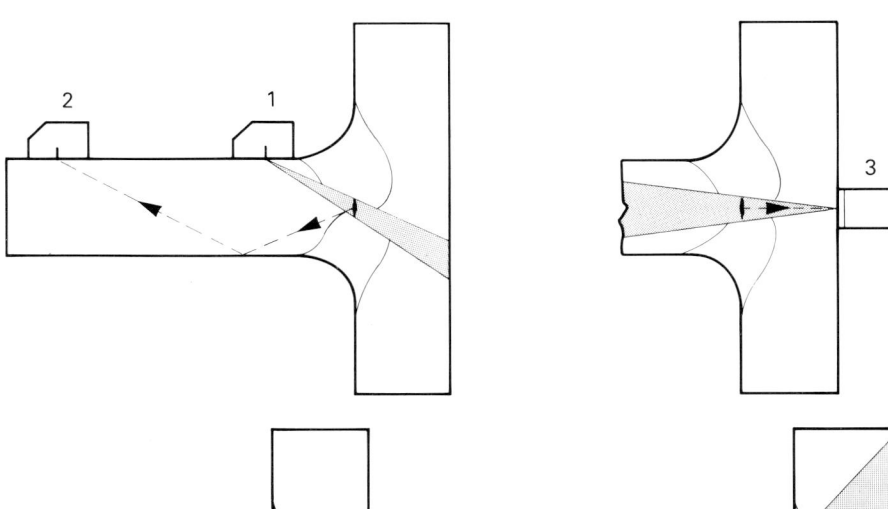

58 Examination of T-weld for HAZ cracking using 45° probes:
A   lower bead;
B   upper bead.

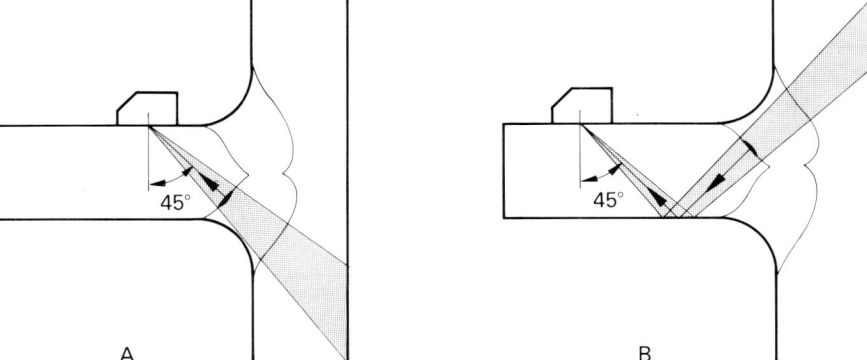

For application to curved surfaces the probe surface can be adapted to match the curvature of the weld and if this is done, the probe index and probe angle must be checked. With a shaped probe shoe there is the possibility of greater beam spread as well as extra mode conversions.

Special techniques

In a short survey of manual ultrasonic testing it is not possible to cover all details of procedure, nor is it possible to take account of all advances in a still developing technique. However, a few potentially important developments are outlined, with appropriate references. Most of these developments are aimed at obtaining more reliable data on defect sizes.

At present the transducer crystal is activated by a short electrical pulse of peak voltage around 500V, but it is possible to excite the crystal with a definite band width which is exactly converted into an equivalent acoustic pulse. This 'narrow-band' transducer will produce an improvement in signal-to-noise ratio, and, it is claimed, much improved results on welds in noisy materials.[13]

An ultrasonic beam can be focused either by using a curved piezoelectric disc or by using a variable thickness disc of some material such as Perspex on the front of the probe; the latter method is easily applicable to compressional wave probes but can also be applied to shear wave probes. Large immersion-coupled focused probes have been used[14] which focus the beam width down to 2mm diameter for a 4MHz shear wave. This means that the specimen is being scanned with a narrower beam of ultrasound, which is equivalent to better resolution of defect edges, etc. A disadvantage of most of these focused probes is that they are restricted to operating at a short depth range, as the ultrasonic beam diverges again beyond the focal point.

A further possibility is to produce a focused beam by a phased array of piezoelectric elements; the focusing system could then be dynamic, to give a variation in depth range.

There are methods of generating ultrasonic waves by systems other than piezoelectric crystals, and some of these are non-contact methods. A special type of probe known as the electromagnetic-acoustic (EMAT) probe is coming into use for applications where coupling to a surface with oil or grease is difficult or impossible, such as on hot steel welds. This type of probe can stand off the surface and generate either compressional or shear waves in the specimen. The EMA transducer probe uses the eddy currents generated by a small coil carrying a high frequency current in the presence of a magnetic field from an electromagnet to generate pulses of ultrasound.[16] There are also methods of generating pulses of ultrasound from a high power laser which allow the beam direction to be controlled.[17]

Ultrasonic surface waves (Rayleigh waves) will follow the faces of a surface-breaking crack so that, if the transit time between two surface wave probes on either side of a crack is measured, the height of the crack is obtained. If the crack is oblique, the crack height and not the depth of penetration into the metal is measured.

Spectrum analysis of the reflected ultrasonic pulse has already been mentioned as a possible technique. At present, the shape of the pulse is dependent more upon the probe characteristics than the defect, but in theory, at least, some information on the defect characteristics ought to be obtainable from pulse spectroscopy.[18]

A high power low noise compressional wave probe can be used to detect the tip of a crack 'end-on' (Fig.59) and can sometimes provide supplementary data on crack size.

The use of ultrasonic diffraction, as in the 'time-of-flight' (TOFD) technique, is relatively new.[19] Ultrasonic diffraction occurs when the beam meets, for example, the tip of a crack (Fig.60), and theoretically there will be diffracted waves from both the upper and lower tips of the crack. If these can be detected and the time-of-flight from transmitter to receiver probe is measured sufficiently accurately, a measure of the locations of the top and bottom of the crack can be obtained.

A technique known as 'Delta-scan' has been used to obtain information about end-on cracks. This is illustrated in Fig.61 and makes use of ultrasonic mode conversion. The transmission probe is on the right and at such an angle that shear waves are refracted into the workpiece at about 60°. If these waves are incident on a vertical crack they are at the correct angle for conversion to longitudinal waves, which travel vertically downwards to be reflected from the bottom of the specimen back into the left-hand receiving probe. Raising or lowering the transmitter probe scans the specimen thickness.

59 Detection of crack tip with compressional wave probe — (T − x) is crack height:
A   probe on specimen;
B   CRT display.

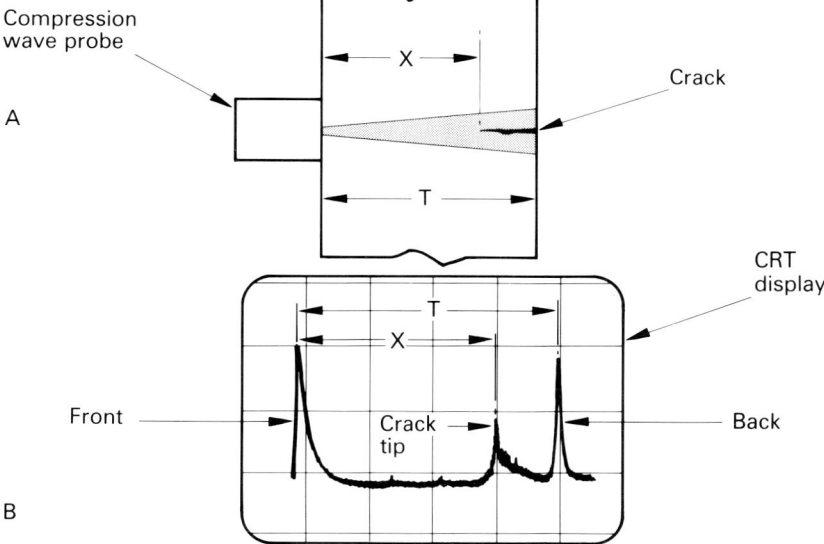

60 'Time-of-flight' crack depth measurement.

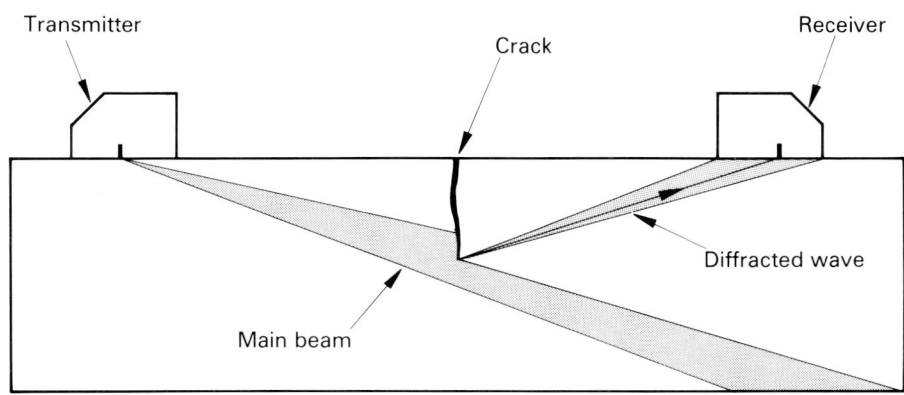

61 'Delta-scan' technique.

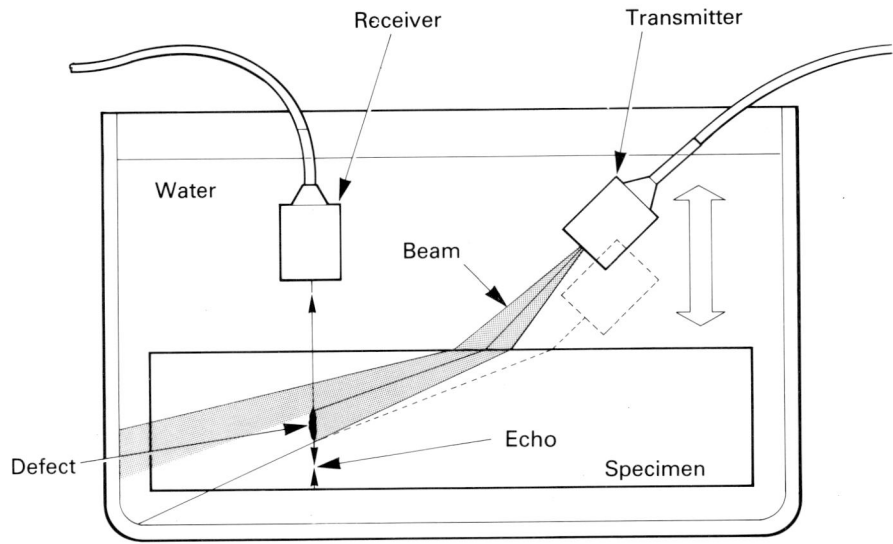

Automated systems

Manually assisted

To ease the problem of interpreting A-scan displays, if the probe position is measured automatically during a scan, two other displays (known as B-scan and C-scan) are possible. In B-scan, the top and bottom surfaces of the specimen are shown as bright lines across the screen, and the defects are shown as bright spots or areas at their correct location through the thickness. The display then is somewhat analogous to a section cut through the specimen. In C-scan, as the probe is scanned on a line raster over the specimen, the display is built up line by line to produce an area image analogous to a radiograph of the specimen over the scanned area. Generally, this display is only used with 0° compressional wave probes and has no large application to weld inspection. It is normally used when the specimen is under water in an immersion tank with a stand-off probe, but can also be used with a contact probe which is connected to two positional potentiometers.[20]

A system called P-scan has also been developed which can be applied to both manual and mechanically scanned probes.[21] In P-scans, the ultrasonic echoes are digitised by a transient recorder and stored together with digitised information on probe position. The associated computer then constructs any form of display that is required (A-, B-, C-scan), as well as presenting digital data on echo heights, etc. The P-scan images are built up on the display screen as the probe is scanned, with appropriate positional markers. This system probably represents the best way of collecting and presenting all the ultrasonic data from a scanning probe; the data can be stored on tape, and hard copies of the display screen can be produced. It has been adapted to mechanical scanning with two probes, one on each side of the weld.

Mechanical scanning

Instead of moving the ultrasonic probe by hand, it was an obvious development to arrange for the probe movement to be done mechanically. To overcome the problem that a defect indication may occur anywhere on the display trace from the input echo to the position corresponding to the maximum ultrasonic path length, this distance can be divided electronically by a series of 'gates'. Any pulse occurring within a gate can then be extracted as an output signal proportional in intensity to the pulse amplitude and used to activate a pen recorder. The first automatic equipments used multichannel recorders.

Instead of attempting to move a probe at right angles to the weld seam between the half- and full-skip distances, it is possible to use a series of probes offset longitudinally from one another, and at different distances from the weld, so that the full weld thickness is covered ultrasonically. To some extent the use of several probes compensates for the loss of flexibility found with a manual system. (When a flaw is detected with a manually held probe, the probe can be rotated slightly to obtain a maximum echo response. This facility is lost with an automatic scanning system.)

To maintain constant coupling conditions it is useful to establish a constant fixed clearance between the probe and the parent plate. This clearance gap is filled with water, continuously, throughout inspection. Travelling speeds depend on ultrasonic path length and pulse repetition frequency, but can be up to 2 m/min. One of the most important practical problems is to have some confirmation that the apparatus is operating properly, that the ultrasonic beams are entering the specimen, and that a zero signal on the recording trace really means no defect signal and not a malfunction such as loss of probe contact, loss of exciting pulse, etc.

62 Two methods of immersion scanning of tubes, using a focused beam probe:
A   beam entering tube at an angle, to detect radial flaws;
B   normal beam, to detect circumferential flaws.

If the weldment is such that it can be placed under water, immersion techniques in a water tank greatly simplify the mechanical problems and can eliminate probe coupling difficulties. The probe is used at a stand-off distance from the specimen, and the display time base is adjusted so that only the ultrasonic path length in the specimen is displayed. The water path length can be adjusted from the simple formula:

$$\frac{\text{Near field length (steel)}}{\text{Near field length (water)}} = \frac{\text{velocity (water)}}{\text{velocity (steel)}}$$

so that the near field of the probe is entirely in the water. Figure 62 shows an application of this type to tube welds.

The mechanical stability and accuracy of construction of the equipment, whether immersion or contact probe, is very important. As in manual testing, a difference in echo distance of 2—3mm can mean the difference between a weld bead edge or backing ring echo and a root crack echo. The design of automatic and remote control ultrasonic equipment has been greatly boosted by the need for equipment for periodic in-service ultrasonic inspection of nuclear reactor pressure vessels. Because of induced radioactivity hazards such equipment has to be installed from a gantry over the reactor vessel, which is water-filled to reduce the radiation level to manageable values. Thus, the ultrasonic probes are several metres from the operator and must be accurately located and manipulated by remote control from the gantry.

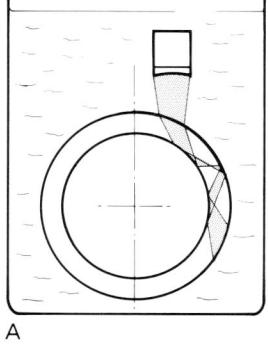

A

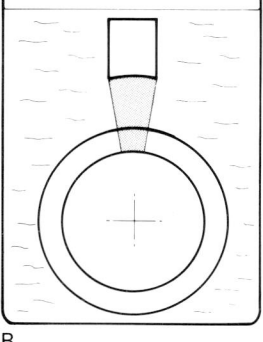

B

Automatic scanning systems with an analogue display are nowadays being replaced by computer-controlled systems in which the electrical signals are converted to digital data, and in which data-processing techniques can be incorporated. Pattern recognition techniques using, for example, adaptive learning networks[22] and spectrum analysis of the pulses are being developed and can, notionally, be applied to ultrasonic data. The present use of these advanced methods is being restricted by the lack of standardisation in probe performance.

Ultrasonic holography, in which the ultrasonic beam from the flaws is mixed with a reference beam, has been successfully developed, but has not, so far as is known, been applied to weld inspection.

Inspection of austenitic steel welds

Austenitic stainless steel is an important material in the chemical and atomic energy industry, and welded fabrications are extensively used. Ultrasonic inspection of such welds is difficult because of the microstructure, but much effort has gone into the development of viable techniques. During cooling, after welding, large grains with a high degree of orientation may develop in the form of large elongated columnar crystals with a fibre texture. This means that the ultrasonic velocity is different in different directions and also that austenitic materials exhibit much greater scattering effects. The general effects of attempting to apply ultrasonic flaw detection to austenitic steel welds are therefore:

(a)   the much higher attenuation, leading to loss of echo pulses;
(b)   much greater grass (noise) due to scattering;
(c)   large spurious signals, due either to grain boundary reflections or to beam bending;
(d)   changes in beam shape and beam path direction.

Minimum attenuation occurs at 45° to the columnar grain axis so that, if weld joints are designed to permit ultrasonic inspection at this angle, the difficulties of inspection are eased. Scattering problems are less at lower frequencies, but this increases the size of the minimum detectable flaw. Short pulse (broadband) transducers and focused beams are claimed to improve results and signal processing techniques can be used to extract signals from noise, but generally the behaviour of ultrasonic waves in austenitic steel weld metal is complex and poses severe inspection problems. Nevertheless, some success has been achieved under specific conditions and there is a considerable literature both on the problems and on applications.[23,24]

Concluding remarks

Ultrasonic weld inspection for the detection of flaws can be applied to most ferritic materials and to light alloys, but only with considerable difficulty to welds in austenitic steels. The equipment needed is not expensive, and unlike radiographic equipment does not increase in price for thick-weld inspection. Because of their simplicity and flexibility manual methods are widely used, but automated systems are also available: these range from relatively simple mechanical scanning to sophisticated computer/digital systems which are likely to be required only for the inspection of very costly plant. All these automatic systems still need guidance on how to collect, compress and interpret data.

For successful manual inspection the key requirement is very accurate calibration of the equipment in terms of beam direction, beam width, etc. For the estimation of flaw sizes there is a range of techniques, none wholly successful and all capable of producing highly erroneous data if not handled with great skill. Many new techniques, different types of probe, etc. are under development, and considerable progress has been made in the standardisation of probes and equipment performance.

Because of the skill required in manual testing and the dependence on the operator's care and integrity, many users of ultrasonic weld inspection require their operators to have certificates of competence in each specific application.

References

1   ESI 98-7: 1982: Ultrasonic probes: normal (0°) compression wave probes for contact testing. Electricity Supply Industry, UK, 1982.

2   ESI 98-8: 1982: Ultrasonic probes: low frequency single crystal shear wave probes. Electricity Supply Industry, UK, 1982.

3   Knott C G: *Phil Mag* 1899 **48** (29) 5.

4   Drury J C: 'Ultrasonic flaw detection for technicians.' Publ Unit Inspection Co Ltd, Swansea, 1978, 26.

5   BS 2704: 1983: Specification for calibration blocks for use in ultrasonic flaw detection. Publ British Standards Institution, 1983.

6   DIN 54 120: 1973: NDT. Calibration block No.1 and its use for the adjustment and control of ultrasonic echo equipment. Deutsches Industries Normes, Berlin, 1973.

7   ISO 2400: 1982: Welds in steel — reference block for the calibration of equipment for ultrasonic examinations. Publ International Standards Organisation, 1982.

8   ISO 7963: 1985: Welds in steel — calibration block No.2 for ultrasonic examination of welds. Publ International Standards Organisation, 1985.

9   'Handbook on the procedures and recommendations for the ultrasonic testing of butt welds.' Publ The Welding Institute, Cambridge, 1971.

10   IIS/IIW 527:77: 'Handbook on the ultrasonic examination of welds.' Publ International Institute of Welding, London, 1977.

11   BS 4331: Parts 1, 2, 3: 1983: Methods of assessing the performance characteristics of ultrasonic flaw detection equipment. Publ British Standards Institution, 1983.

12   BS 3923: Parts 1, 2: 1986: Methods for ultrasonic testing of welds. Publ British Standards Institution, 1986.

13   Deutsch V and Crostack H A: *Brit J NDT* 1980 **22** (4) 166.

14   Saglio R and Prot M: *Materials Evaluation* 1978 **36** (1) 62.

15   Whittingham T A: 'Physical aspects of medical imaging.' Publ John Wiley & Sons, 1981, 153.

16   Whittington K R: *Brit J NDT* 1981 **23** (3) 127.

17   Scruby C B: *Brit J NDT* 1981 **23** (6) 312.

18   Gericke O R: *J Amer Acoustic Assoc* 1962 **35** 364.

19   Silk M: 'Developments in pressure vessel techniques.' edited by R W Nichols. Publ Applied Science Publishers, London, 1979, Ch 4.

20   Lavender J D and Wrightson J C: *J Amer Foundrymans Assoc* 1976 **84** 156.

21   Nielsen N: *Brit J NDT* 1981 **23** (2) 63.

22   Whalen M F and Mucciadi A N: 'NDE in relation to structural integrity.' edited by R W Nichols. Publ Applied Science Publishers, London, 1980, Ch 5.

23   Whittaker J S and Jessop T J: *Brit J NDT* 1981 **23** (6) 293.

24   International Institute of Welding: 'Handbook on the ultrasonic examination of austenitic welds.' Publ American Welding Society, 1986.

4   Magnetic methods

Magnetic particle testing is a cheap and simple method of non-destructively detecting cracks which reach the surface in ferromagnetic materials. Under very favourable conditions it can also detect sub-surface cracks. In addition, there are certain other magnetic testing methods such as magnetography which will be briefly described.

Considering the simplicity of the technique, magnetic particle testing is not applied to weldments as widely as might be expected. It is applied to fillet welds and to welded repairs but could be used much more widely as a preliminary to radiographic and ultrasonic examinations.

Principles

The principle of the method is that the specimen is magnetised so as to produce magnetic lines of force in the material. If these lines of force meet a discontinuity, such as a crack cutting the lines, secondary magnetic poles are produced at the faces of the crack, and if these are near the surface they can be revealed by applying magnetic particles as a powder or in a liquid suspension (Fig.63). Cracks are shown at maximum sensitivity when they are at right angles to the magnetic flux, and when they reach the specimen surface so that there is a local leakage of flux out of the specimen.

The weldment can be magnetised in several different ways. Broadly, the specimen can be made part of a magnetic loop, i.e. by using the specimen as keeper across the poles of a permanent or electromagnet; the specimen can be put inside a coil carrying an electric current; a current can be passed directly through the specimen, or through a bar threaded through the specimen. A very important and useful general rule is that, for all current flow methods, the magnetic lines of force are at right angles to the current direction and, as the best direction for crack detection is to have the crack cutting the magnetic lines at right angles, it follows that the current flow direction should be approximately parallel to the expected direction of the crack. This is a basic rule which governs the choice of a suitable magnetising technique.

Magnetic testing depends upon the difference in magnetic properties between a defect and the main mass of metal. This difference is a maximum with an air-filled crack, but slag, paint, carbon and oil are equally non-magnetic and, if

63 Magnetic particle test indicating longitudinal crack in weld.

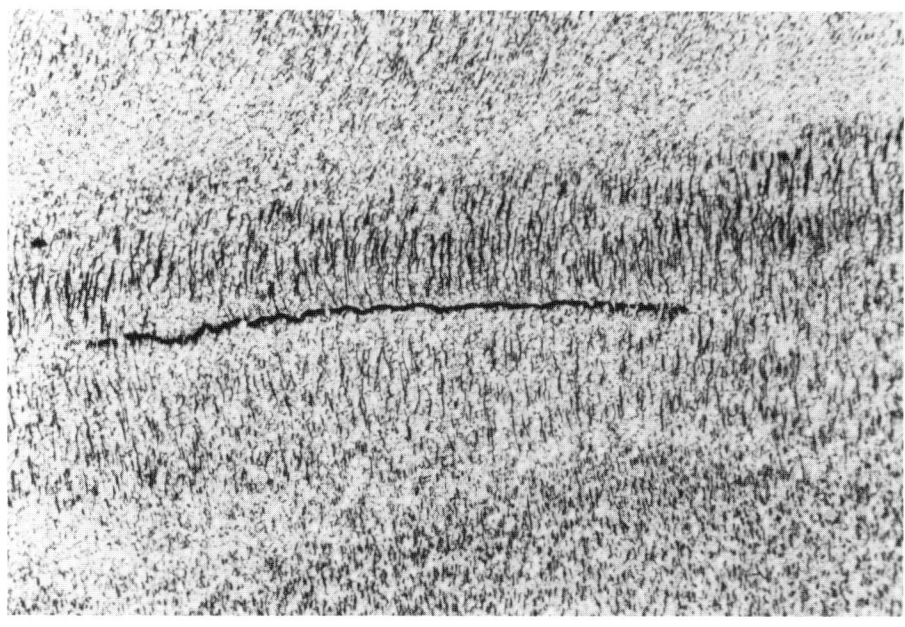

present in a crack, will not prevent the method working. It is not necessary, therefore, to remove contamination from the crack, since a distorted magnetic field resulting from it will still be present at the surface. This field provides a region on the surface which is capable of attracting any mobile magnetic material. The force of attraction depends on many features, which are identified later, but the overall effect is to cause magnetic particles to map out on the surface a pattern which exactly reproduces the shape of the magnetic field. The width of aligned particles may be many times greater than the crack, and this magnification enables cracks down to a few microns wide to be made visible to the naked eye when the contrast of the indication with the surface is good.

The main features of magnetic particle testing are that:

(a) special preparation of the surface to be tested is not necessary: loose paint, rust or scale must be removed, but a thin layer (< 50μm) of adhering paint need not be. It is not necessary for the surface to be ground or polished;

(b) it is essential to be able to induce a magnetic field of the required intensity into the component. Ideally, the field should be uniform in the region under inspection, but frequently this is impossible because of the shape of the component. Limitations to inspection resulting from this factor must be appreciated, and are discussed in detail below;

(c) the media used to detect the distortion of the magnetic field at the surface, caused by flaws, are of paramount importance. It is obvious that, even if ideal magnetising conditions have been produced, they will be useless unless the indicating media are satisfactory.

The limitations of the technique are that:

(a) it cannot be used on non-magnetic materials;

(b) components of complex shape may require several tests to ensure complete coverage;

(c) on complex shapes, variations in magnetisation may produce variations in flaw sensitivity.

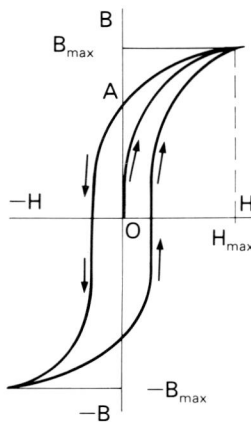

64 Hysteresis curve (B against H) for a ferromagnetic material. OA is the remanent magnetisation when the magnetising field first returns to zero.

There are large differences in the magnetic properties of steels, which result in very different magnetisation (B—H) curves, B being the magnetic flux density and H the magnetising field strength. The value of the magnetic flux density produced depends on the permeability, μ, of the material ($\mu=B/H$), which rises from zero to a maximum value and then falls (Fig.64). If the magnetising field is increased to H and then decreased, the magnetic flux density is not normally zero when H returns to zero (point A) and there is a residual (remanent) magnetisation OA, which is sufficient in some materials for magnetic particle testing (see next section). For crack detection it is best to magnetise the specimen with a magnetising field slightly greater than that required to give maximum permeability, but many factors need to be considered in deciding on the best magnetising procedure.

Methods of magnetisation

The basic principle is to produce magnetic lines of force *across* the expected direction of any cracks. If the likely crack direction is unknown, the test must be performed in two directions at right angles, or one of the more sophisticated methods using a rotating or double field must be used.

The basic magnetisation methods are:

Magnetic flow: to make the specimen part of a magnetic circuit by effectively using it as the bridge of a permanent or electromagnet; the latter method is sometimes called the magnetic yoke technique (Fig.65).

Current flow: to pass an electric current through the specimen, broadly along the direction and through the region in which cracks are to be expected (Fig.66).

Induced current flow: used for ring specimens, by effectively making them the secondary of a mains transformer (Fig.67).

Electromagnetic induction: to pass an electric current through a conductor which is threaded through a hollow specimen (Fig.68) or placed adjacent to or wrapped around it (Fig.69). A relatively new variant of this technique is to have a current-carrying flat coil laid on the surface of the specimen, with small spacers to hold it off the surface.

65 Magnetic yoke technique.

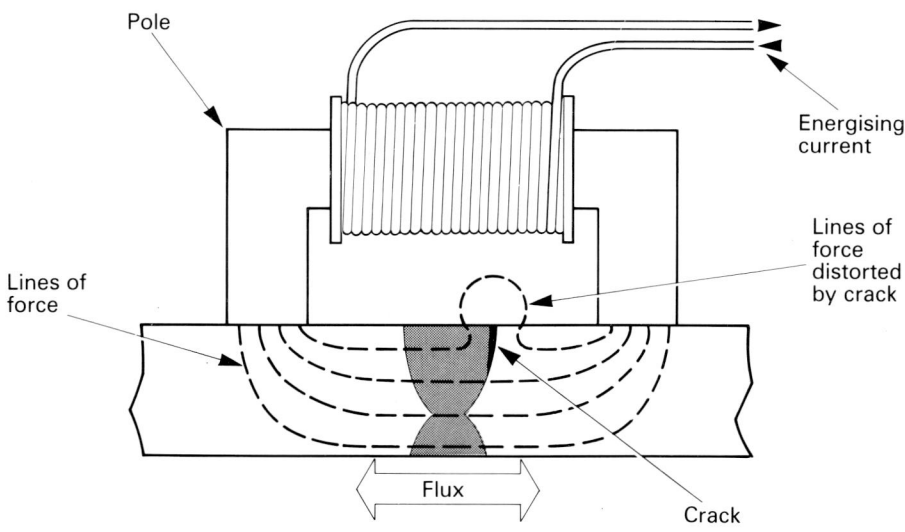

66 Current flow technique.

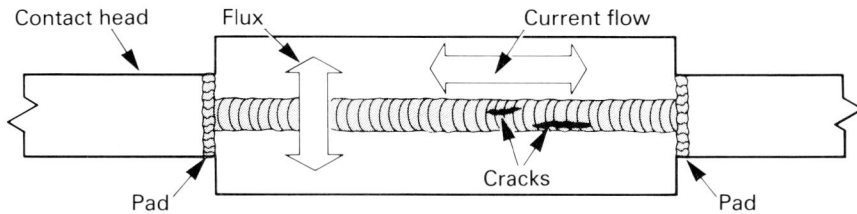

67 Induced current flow applied to a ring-shaped specimen.

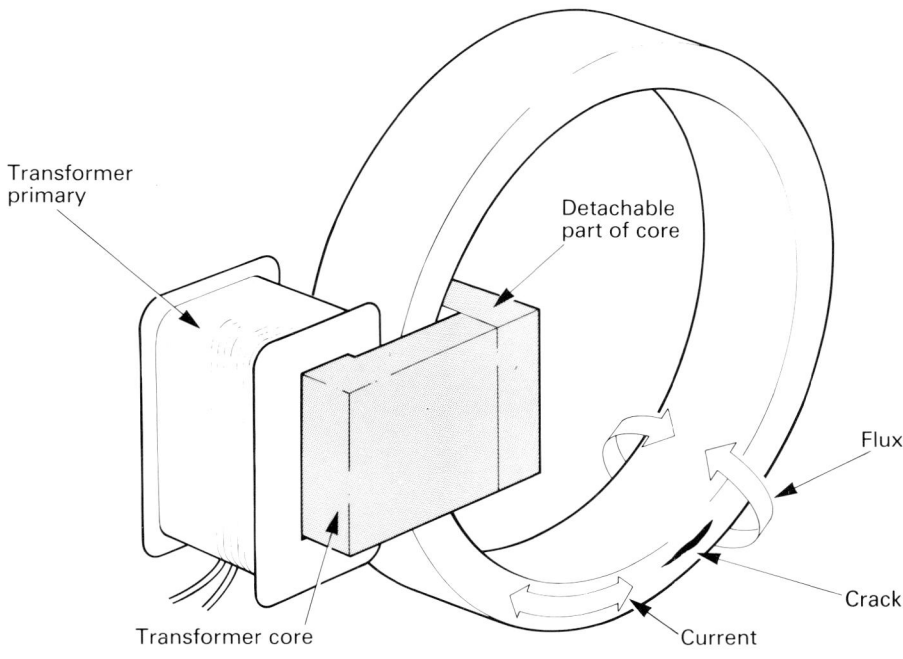

68 Electromagnetic induction technique I: current-carrying lead threaded through hollow section.

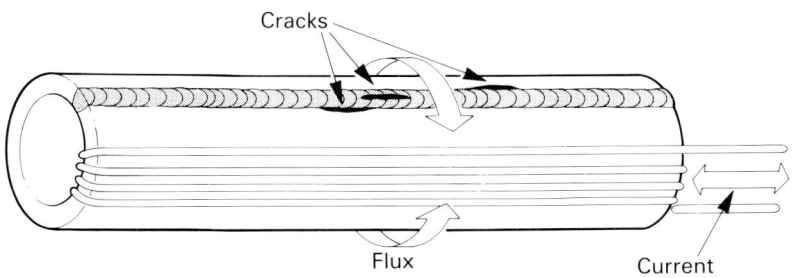

69 Electromagnetic induction technique II: current-carrying lead wrapped around outside of workpiece.

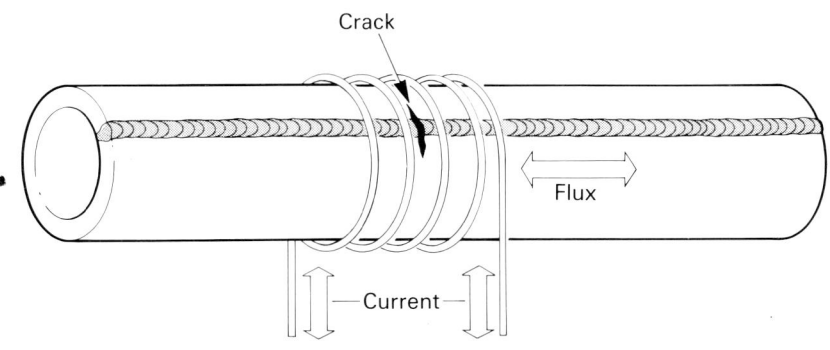

The choice of magnetising technique depends chiefly upon specimen geometry, and for large welds in plates or piping the current flow method, in which the current is applied by means of prods to local areas of the weld is usually likely to be the most convenient.

A relatively weak magnetising field will suffice for magnetic crack detection when the detector (ink or powder) is applied during magnetisation, and even when the remanent magnetism is used after momentary application of a field, most ferromagnetic materials contain sufficient residual magnetism to display cracks. Nevertheless, it is obviously desirable to use magnetising flux densities which will produce optimum results. The recommended magnetic flux density according to BS 6072: 1981 is 0.72 tesla.[1] (The tesla is the SI unit of magnetic flux density, defined as 1 weber of magnetic flux per m^2.)[2]

If residual (remanent) magnetism is to be used, after alternating current application, the circuit breakers must be of a type which ensures that the current is not broken on the rising part of the half-cycle, where the value of B (Fig.64) could be zero. It is still a point of controversy as to what type of circuit breaker is necessary.

Magnets and DC electromagnets

There are a number of commercial patterns of these in which the pole faces are flexible so as to ensure better contact on to the specimen than a flat rigid face. According to BS: 6072: 1981, suitable magnets should have a lifting power of 18kg and a pole spacing of between 75 and 150mm. Because such magnets still provide only a low level of magnetisation, it is recommended that they should be used only when no other techniques of magnetisation are practical.

Current flow using prods

The current from an external source such as low voltage transformer winding is passed between two prods held on to the specimen surface (Fig.70) and the crack to be detected should lie within ± 45° of the direction of current flow. If practical, contact clamps are better than hand-held prods. Typical prod spacing is 200mm, with a peak current of 700-1000A.

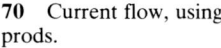

70 Current flow, using prods.

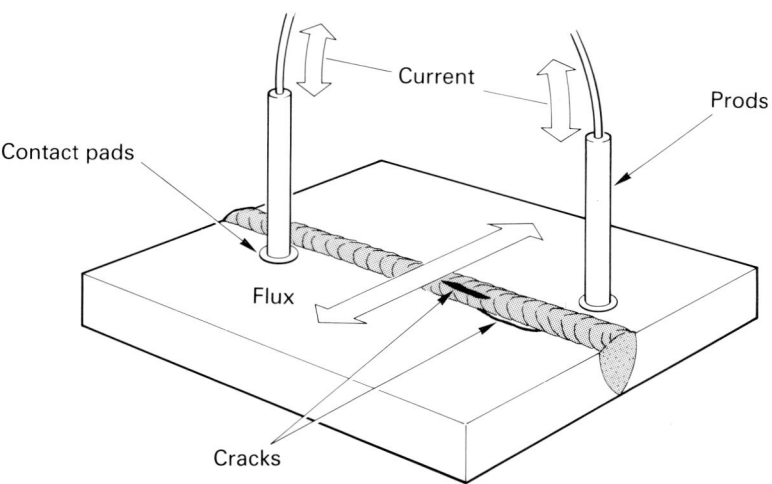

Precautions must be taken to prevent burning and arcing at the contact points, usually by using contact pads of mesh or braid, made of steel or aluminium. Lead caps on the prods are also used but are not recommended because of the production of noxious gases from the lead. Copper prods and copper mesh should be used with caution, as copper can cause metallurgical damage. The correct technique is that the current is not switched on until adequate contact pressure has been established, and the contact pressure is not released until the current is switched off.

Threaded bar technique

When a tubular component is threaded on to a single bar, currents of 5-7 A/mm diameter of the component are required (Fig.71). If a flexible cable is used instead of a bar, the current required is reduced in proportion to the number of turns. The direction of the magnetic flux is around the circumference of the hollow component, so the method is used to detect only the longitudinal ($\pm 45°$) cracks. However, if the flexible current-carrying cable is wound circumferentially around the outside of a tubular component, circumferential cracks can be detected.

Other methods

There are several other possible magnetising techniques but these are not likely to be applied to welded fabrications. More sophisticated magnetising methods, using combined AC/DC or phase-shifted double AC magnetising currents, are claimed to eliminate the need for inspection in two directions and to give much improved results on irregular section specimens. Whether such methods are necessary depends upon the quantity of work envisaged.

Magnetising current

The magnetising current used and the method of measuring it affect the optimum value quoted. According to BS 6072: 1981 if a direct current is given a factor of 1.00, the indicated current values should be multiplied by the factors shown in Table 7 when a moving iron electrodynamic or induction ammeter is used for the measurement of the current.

The relative merits of AC and DC magnetising currents are still the subject of discussion, with most of the technical arguments seeming to point in favour of AC, although the equipment is likely to be slightly more expensive. If AC is used, the current flow is close to the surface (the skin effect), and with variable

71 Threaded bar technique for hollow sections.

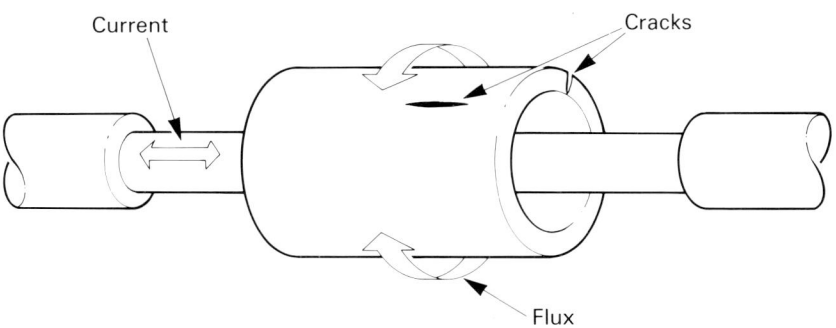

Table 7. *Factors by which indicated current values should be multiplied when using a moving iron electrodynamic or induction ammeter*

	AC	Full-wave rectified single-phase	Half-wave rectified single-phase	Full-wave rectified three-phase	Half-wave rectified three-phase
Factor:	1.00	1.41	2.00	1.05	1.19

cross-section specimens there is a better chance of obtaining adequate magnetic flux over the whole surface. With DC, the cross-section of the specimen is filled with smooth flow lines which do not follow the surface, so that local regions of greater cross-section may not be adequately magnetised.

Magnetic field intensity indicators

It is rarely practical to measure the magnetic field directly, but portable flux indicators can be used to confirm the field direction and provide a rough guide to flux levels. One suitable commercial indicator consists of a 50×12mm strip of Mu-metal containing a narrow slit, encased in thin copper, which is used by fixing it to the specimen surface with transparent adhesive tape, correctly aligned. After the magnetising field is switched on and flaw indication ink applied, the indicator should show the line of the slit in the Mu-metal if the magnetising field is adequate.

Detection media

During or after magnetisation a 'magnetic ink' or powder is applied to the magnetised area of the specimen to reveal any cracks. Magnetic inks are available in both fluorescent and non-fluorescent forms and consist essentially of finely divided ferromagnetic particles suspended in a suitable liquid carrier. Early inks were always black and the carrier was a light paraffin, but nowadays different colours are available and the carrier may be water.

Magnetic powders also consist of finely divided ferromagnetic particles. There are specifications (e.g. BS 4069: 1982[2]) for the solid content and foreign content, and settlement time of magnetic inks and functioning tests, and different grades of ink are marketed. Magnetic inks contain a range of particle sizes which is an important part of their constitution. The ink must be very thoroughly mixed before use.

In the UK, most applications of magnetic particle crack detection use magnetic ink, but in the USA powder is much more widely used. There is no obvious technical reason for these different preferences. After the ink or powder is applied, sufficient time must be allowed for indications to build up: the ink or powder must be used liberally and applied gently so as not to wash away any indications. The ideal is that the ink flows over the surface with very little pressure, so that the particles can accumulate at a flaw indication without being washed away. Aerosol spray applicators should be used with great care. For small specimens, an immersion procedure is a good and convenient method. It is important that ink application ceases before the magnetising current is switched off.

After application of the ink, the component should be allowed to drain for a few seconds; this usually increases the contrast of the indications.

Magnetic powders are considered more suitable than inks for use on very rough surfaces, on hot surfaces and to detect sub-surface cracks. Again, they must be applied very gently, being made to float on to the surface as gently as possible and not be thrown on. Specially designed low-velocity powder blowers are available to apply magnetic powder.

Depending upon the colour of the specimen surface, either coloured or fluorescent inks or powder may be necessary to produce good contrast images.[3] Alternatively, the surface can be coated with a quick-drying white contrast paint before application of the ink or powder. If fluorescent inks are used, ultraviolet (UV-A) radiation for viewing must be provided and the ambient room lighting reduced, preferably below 10 lux. For all other conditions good lighting for viewing the indications is necessary, preferably not less than 500 lux, e.g. a 100W pearl bulb at 200mm distance.

For large structures, examined *in situ*, the use of a white contrast paint and black ink is fairly common on unmachined surfaces. Such a paint may be an acetone/zinc oxide/cellulose binder, which dries instantly and is not particularly affected by oily surfaces. The acetone vapour may be objectionable in restricted areas; tin oxide can be used instead of zinc oxide. For larger areas, ordinary white emulsion paint can be used; this dries in 1-2min but the surfaces need to be grease-free before application. Whichever contrast paint is used the coating should be kept very thin, and if the specimen is subsequently to be heat treated the paint must be brushed off after inspection.

Inspection procedure

The pre-inspection procedure necessary on a welded surface is minimal. Areas to be examined should be free from scale and dirt, largely to ensure that the ink does not become contaminated: this is particularly important if fluorescent inks are to be used. Specimens should be dry, as water in the paraffin-based inks can cause flocculation.

If current-flow prods are to be used, the contact points for the prods should be cleaned locally and the plan of inspection to cover an area decided upon. If a contrast paint is to be used, this is now applied between the prod contact points and a field intensity indicator is taped on. Assuming that the ink applicator is loaded and ready, the magnetising current can now be applied. Some ink is flooded over the area before the current is switched on, and more is applied while the current is flowing. (If the residual magnetism technique is being used, application of ink can be left until after the magnetisation.) After an interval of 10sec the magnetised area can be inspected.

Interpretation

Crack-like flaws are the ones most easily detected and are shown as crisp sharply defined lines; inclusions or slag stringers give more diffuse indications. It is impossible to draw any conclusions on the height or width of a crack from the magnetic indications on the surface. False indications can occur where there is a section change or a change in magnetic properties, or because of tool marks or scratches which hold the magnetic ink. Magnetic writing caused by two specimens rubbing over one another can also cause spurious marks in certain steels. These false indications are seldom troublesome; if necessary, local inspection at low magnification will usually enable the inspector to see the crack and distinguish cracks from scratches.

Crack indications can be preserved with adhesive tape, or strippable coatings, or with special proprietary techniques such as magnetic rubber, but these methods are not widely used. Photographic records can also be made of the indications.

Demagnetisation

Steel articles which have been exposed to the effects of very strong magnetic fields often remain magnetised for a considerable time after testing. This is troublesome when the component is built into machinery, because local poles attract ferrous particles which may cause excessive wear. On heat treatment, demagnetisation takes place automatically, but for the majority of components tested in the final-machined condition demagnetisation methods have to be used.

With DC demagnetisation, the current is repeatedly reversed while being progressively reduced. With AC demagnetisation, which is the most common method, the component is subjected to a gradually diminishing field by either withdrawing it along the axis of a solenoid or reducing the current in the solenoid. Better results are obtained if the part is rotated as it passes through the coil. A special technique to demagnetise large structures is to pass a special low frequency AC through the part; however, only DC is suitable for very thick section parts, since AC will demagnetise only the skin. The parts should be checked for freedom from magnetism after treatment.

Magnetographic methods

Instead of applying magnetic ink or powder to the specimen, it is possible to have a strip of magnetic tape laid along the weld surface during magnetisation; any variations in the surface magnetic flux will be recorded on this tape. The tape can then be removed and examined by scanning along it with a magnetic head (coil or Hall-effect probe), amplifying and processing the indications, and presenting the results on a measuring instrument or on an oscilloscope display. The method has been known for several years and appears to have considerable potential, but reports of successful applications are very few. Special thick and heavily loaded magnetic tape is necessary for good results.

Sub-surface crack detection

The conditions under which sub-surface cracks can be reliably detected are very restricted, and magnetic particle inspection should not be considered as suitable for such applications.

If the surface of the weld can be ground to a very smooth finish, a very high DC magnetising current used, and powder rather than ink employed as the detecting medium, sub-surface cracks can be found, even when these are up to 10mm below the surface. The technique has been applied to circumferential pipe welds. However, the cost of surface preparation makes the use of other inspection methods, such as ultrasound, economically preferable and potentially more reliable.

References

1   BS 6072: 1981: Method for magnetic particle flaw detection. Publ. British Standards Institution, 1981.
2   BS 4069: 1982: Specification for magnetic inks and powders. Publ. British Standards Institution, 1982.
3   BS 5044: 1982: Contrast aid paints used in magnetic particle flaw detection. Publ. British Standards Institution, 1982.

General background reading

Blitz J, King W G and Rogers D G: 'Electrical, magnetic and visual methods of testing materials.' Publ Butterworth, London, 1969.

PD 6513: 1985: Magnetic particle flaw detection: a guide to the principles and practice of applying magnetic particle flaw detection in accordance with BS 6072. Publ British Standards Institution, 1985.

BS: 3683: Parts 1-5: 1983-85. Glossary of terms used in non-destructive testing. Publ British Standards Institution, 1983-85.

5   Penetrant methods

General principles

Penetrant methods comprise a range of techniques in which a liquid is put on the surface of the specimen and given time to soak into surface-breaking cracks and cavities. The surplus liquid is then removed from the surface and any liquid which has entered cracks, etc. is made visible by a developer, fluorescence, or seepage (Fig.72). Normally penetrants are applied to one surface, but leakage defects can be found by application on one side and testing for traces of penetrant on the other. In principle, penetrant methods can be applied to all weldments to detect surface-breaking cracks, but in practice magnetic particle methods are preferred on welds in any material which is magnetisable. Penetrant testing is widely applied to welds in light alloys, austenitic and non-magnetic steels, and other non-ferrous materials. It has the inherent advantage that it can be applied to the whole surface area of a specimen in one operation. There are two important classes of penetrant:

· *Fluorescent* materials, with which a source of ultraviolet (UV-A) light is needed for viewing;
· *Dye* penetrants, which give a coloured indication, usually red, on a white background (Fig.73).

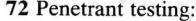
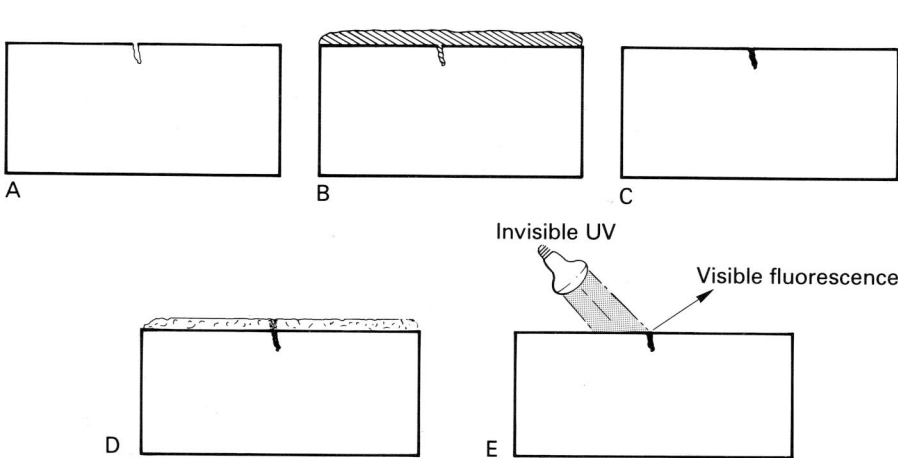

72 Penetrant testing:
A sample before testing;
B dye applied, penetrates into cracks;
C surplus wiped off leaving penetrant in crack;
then either:
D developer powder applied, dye soaks into powder;
  or
E UV lamp shows up fluorescent penetrant.

Invisible UV

Visible fluorescence

73 Surface crack
detection with dye
penetrant.

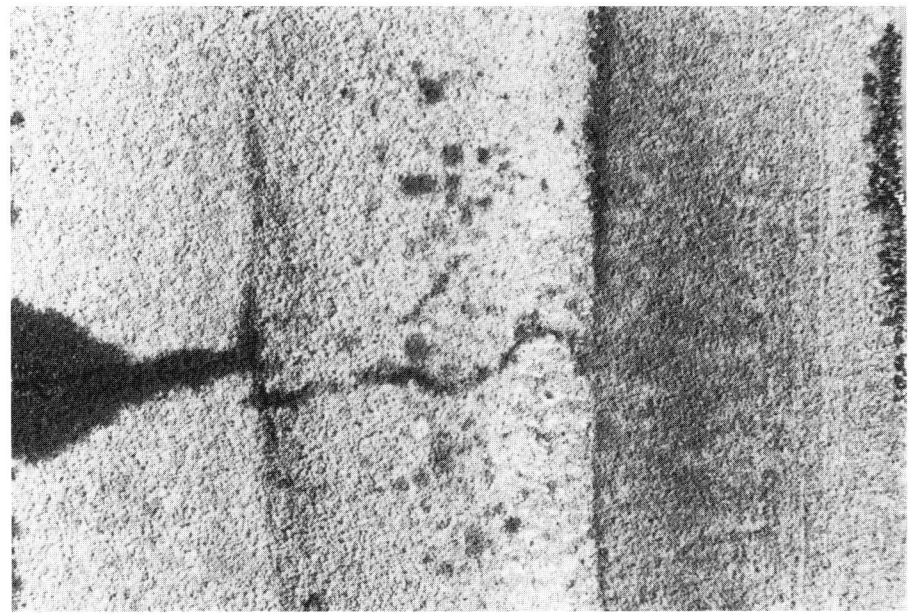

Fluorescent penetrants can be further sub-divided into those which deposit a material into the crack which fluoresces when dry, and a second type consisting of fluorescent liquids.

In some materials a double dye is used in which fluorescent light from the first dye activates fluorescence in the second when UV light is applied, so producing a greater brilliance than with a single dye. The UV light used with fluorescent penetrants (wavelength around 360nm) is often referred to as 'black light'.

Penetrants, as the name implies, must be able to penetrate into narrow cracks but not be easily removable from wide or shallow discontinuities until required. They need to have an intense colour or fluorescence. The key to success with any penetrant testing is very thorough pre-cleaning. If a crack is full of some foreign material such as oil or water, penetrants cannot enter and the crack will not be revealed. One of the most effective cleaning methods is the use of trichlorethylene vapour, or alternatively swabbing or immersion in a solvent. For small specimens, ultrasonic cleaning baths are very useful. Abrasive methods of cleaning should not be used, unless they can be followed by an acid etch, as they tend to close over the mouth of a crack.

Penetrant systems have been developed for *in situ* inspections on overhead or vertical surfaces, using a thixotropic (jelly-like) penetrant liquid. There is said to be no loss in sensitivity compared with conventional liquids when both are used on a horizontal surface.

Types of penetrant process

Dye penetrants

These are available in three versions:

(a)   solvent-clean (solvent removable);

(b)   water wash (sometimes called pre-emulsifiable);
(c)   post-emulsifiable.

For the solvent-clean type, the sequence of operations is: pre-clean; apply penetrant; remove excess by wiping surface with dry cloth or paper damped with solvent remover; apply developer; rapid drying; inspection.

For the water wash type, the surplus penetrant is removed by spray rinsing with water, and the specimen is dried either with compressed air, or in an air-circulating oven.

The post-emulsifiable type of penetrant requires an extra stage in which application of a detergent type of penetrant remover is followed by water wash and drying. There are two types of penetrant remover:

(a)   lipophilic: which depends on mutual solubility between it and the penetrant, so that contact times must be controlled to a minimum value;
(b)   hydrophilic: which is not soluble in the penetrant, so that the penetrant remaining in the cracks is not removed. Hydrophilic removers are water-tolerant, so that excess penetrant can be removed by water washing prior to the application of remover.

Dye penetrants are simple to use, are supplied in aerosol kits, and are particularly suitable for field work as no UV lighting is needed. They are generally more costly than fluorescent materials and do not achieve as high a contrast of indication against background.

Fluorescent penetrants

These are either:

(a)   water-washable;
(b)   post-emulsifiable.

The procedures are exactly the same as for the same types of dye penetrant with the emphasis again on thorough pre-cleaning and removal of any chemicals such as chromates or chromic acid before applying the penetrant material. A typical penetrant contact time is 20-30min, but some manufacturers claim that shorter times can be used.

The washing operation can advantageously be carried out under UV illumination so that it can be terminated as soon as the specimen is clean. Either dry powder or a wet developer (water suspension) can be used with the water-washable process, and if the latter is used the drying stage is omitted. If dry powder is used, the specimen must be thoroughly dry and preferably warm (65°-80°C) as this causes the penetrant to exude. A typical 'developing time' with dry powder is 20min.

With fluorescent penetrants the inspection must be under UV light in a darkened area. A completely dark room is not essential and may be disadvantageous from the handling and identification point of view. The sources of UV light (Fig.74) should be positioned so as to avoid the eyes of the inspector. The difference between the water-washable and post-emulsifier fluorescent processes is the same as for dye penetrants, in that the application of the emulsifier is a separate stage which enables the degree of washability to be controlled by controlling the emulsifier/penetrant contact time. Generally, post-emulsifier processes, although they contain the extra stage and need careful timing, are considered to be capable of achieving higher sensitivity.

74 Ultraviolet (UV) lamp to provide localised illumination for fluorescent dye penetrant testing.

Special techniques

For small components the use of post-emulsifiable penetrants has enabled the process to be automated, so standardising the processing cycles. Normally, the components are loaded into wire baskets which are attached to a conveyor rail; the process is more applicable to small components such as turbine blades and castings than to weldments. The total process time, including drying and development, is likely to be about one hour. Both dye and fluorescent penetrants are used, but the latter is more common because of its greater sensitivity.

The advantage of an automated process, as applied to new components having a known history, is that the preliminary pre-cleaning process, which is the key to reliability in penetrant inspection, can be matched to the contaminants known to have been used during manufacture. The major problem in penetrant inspection is the pre-cleaning of a specimen of unknown history.

Some dye penetrant processes can be applied at high temperatures and there are important applications in the examination of preheated welds, part-completed welds, and weldments which require a very slow cooling rate. The main differences in procedure are in the soaking temperatures, which are reduced from 20min at 20°C to $1\frac{1}{2}$min at 100°C or 30sec at 150°C.

Sensitivity

Ever since the introduction of penetrant examination, attempts have been made to devise methods of comparison of the different penetrant processes. One type of test specimen was a demountable block in which annular mating surfaces, previously lapped together, were locally relieved by the removal of 2-10μm.

These faces were then tightened together by a central bolt. The narrow discontinuities thus formed are not natural cracks, and the lapped faces do not resemble the faces of a natural crack. It was found that the narrowest gap (2μm) was much too large to differentiate between different penetrants. The only advantage of an artificial test piece of this type is that it can be dismantled and cleaned properly before re-use.

A second type of penetrant test piece has been made by developing a pattern of natural cracks in a standardised block of 2024 aluminium. By a standardised heat/quench treatment it is claimed that cracks of reproducible width and depth are produced. Each block is divided into two halves for comparison tests. After use, the blocks can be cleaned by vapour degreasing and stored in acetone.

Another type of test block consists of a strip of steel which has been chromium plated to a thickness of about 100μm. Under appropriate plating conditions the plating develops microcracks. This type of test block is available commercially, and the cracks are fine enough to show up clear differences in the performance of commercial penetrant processes. It is claimed that the test blocks can be cleaned and re-used.

The UV light used for penetrant inspection lies in the wavelength range of 320-400nm (UV-A) and is often called 'black light'. There are standard instruments to test the output of these UV lamps, based on a photocell pick-up and a meter (BS 4489: 1984);[1] some use an intermediate fluorescent screen as a converter and some measure the black light intensity directly (Fig.75). Most UV lamps are of the mercury-arc type with a glass filter which removes all visible light and all UV shorter than 300nm. The UV output of these lamps is liable to deteriorate and should be checked at regular intervals.

Sensitivity and reliability depend more on pre-cleaning and surface condition than on any other factors. All forms of grease and oil have an adverse effect, and corrosion products and solid residues such as oxides can completely invalidate the results. The degreased surfaces of ferritic materials should therefore be inspected immediately after degreasing.

Very tight cracks may originally show only as a row of dots, which link up into a continuous line after a period of time, but generally cracks appear as clearly defined lines of high contrast.

In choosing a penetrant system for a particular application it is possibly more helpful to make the choice on the basis of an appropriate removal process for the specimen, rather than directly on sensitivity considerations. The choice of developer also influences the sensitivity of the process, as solvent suspensions offer the greatest potential sensitivity when they can be used, followed by aqueous developers and then dry powders. The third factor is the choice of viewing conditions and whether UV light and suitable viewing conditions for its use are available.

Normally there is no special hazard in penetrant inspection, but for large runs of work precautions against inflammable vapours and dust need to be taken.

References

1   BS 4489: 1984: Method for measurement of UV-A radiation (black light) used in non-destructive testing. Publ British Standards Institution, 1984.

75 UV monitor to check performance of UV lamps in fluorescent dye penetrant testing.

General background reading

M 39: Methods for penetrant testing of aerospace materials and components. Publ British Standards Institution, 1972.

ISO 3879: 1977: Recommended practice for liquid penetrant testing of welded joints. Publ International Standards Organisation, 1977.

ISO 3452: 1984: NDT — penetrant inspection — general principles. Publ International Standards Organisation, 1984.

ISO 3453: 1984: NDT — penetrant inspection — means of verification. Publ International Standards Organisation, 1984.

Blitz J, King W G and Rogers D G: 'Electrical, magnetic and visual methods of testing materials.' Publ Butterworth, London, 1969.

'Non-destructive testing handbook.' 2nd edition. Vol 2: 'Penetrants.' Publ. American Society of Non-Destructive Testing, Columbus, Ohio, 1983.

6   Electrical methods

Included in this chapter are the eddy current and potential drop methods of testing for crack depth measurement, and electrical techniques of metal sorting. Magnetic methods of flaw detection were covered in Chapter 4.

Eddy current methods

Eddy current testing can be carried out on any material which conducts electricity. If a coil carrying an alternating current is placed near a conducting metal specimen, eddy currents are induced in the specimen and in turn produce a current in a second search coil, or affect the current in the primary coil. The effect is analogous to a transformer, with the specimen acting as the transformer core. The flow of eddy currents in the specimen is disturbed or impeded by flaws in the specimen and so the reading of the search coil current is an indication of these flaws. The primary and search coils may be on opposite sides of a thin sheet specimen, may be wound coaxially, or may be separate side-by-side coils. The coils may be annular (encircling) through which tubular specimens are passed, wound on internal bobbins to pass through tubing, or in the form of a single transmitter/receiver coil.

The voltage induced in the pick-up coil, apart from being dependent upon the primary input current, depends upon the stand-off distance of the pick-up coil, the metallurgical condition of the specimen (permeability, conductivity) and its dimensions as well as on the defects in the specimen. If the geometry and the metallurgical state are accurately controlled, eddy current testing can be a very sensitive method of finding flaws, but this is rarely the situation with welds.

By measuring both pick-up coil voltage and impedance at a range of frequencies, effective permeability curves (normalised voltage and impedance) can be built up for any coil geometry and these show that, for certain frequency values, the effect of coil stand-off (or coil fill-factor with an annular coil) can be minimised. The subject is complex, and because it has little application to weld inspection, is not described in detail.

The principal use of eddy current testing is for the automatic and very rapid examination of machined tubes, wires, bar stock, etc. BS 3889: Part 2A: 1986 covers the inspection of ferrous pipes and tubes up to an OD of 60mm. Some eddy current test equipment has been built for surface crack detection using a hand-held search coil which, after appropriate adjustment, is moved slowly across a surface, and it has been claimed that the readings can be taken as a measure of crack depth. Instruments of this type range from very simple portable, single-frequency, battery-operated models, to multifrequency instruments, some having a phase display output, in which the shape of the output curve is claimed to indicate whether the defect is a crack or an inclusion or a surface mark. There is evidence that the multifrequency instruments can, under controlled conditions, measure crack depth successfully, but they are not likely to function successfully on welds unless the surface condition is good, and it is very unlikely that they will function satisfactorily on an as-welded surface.

Eddy current testing on non-ferrous materials is much easier and more satisfactory than on ferrous materials.[2,3]

Potential-drop methods

If a pair of contacts is placed on a metal specimen and a voltage applied across them, a current will flow, and, if the metal between the probes contains a crack transverse to the current direction, the resistance will be slightly higher and less current flows. This is the potential drop method. In practice, the metal is such a good conductor that the changes are very small and almost impossible to measure, but, if a second pair of contacts with a small (fixed) spacing is placed first across the crack and then across sound metal, the two voltage readings on this search probe will be different, and this difference will depend on the height of the crack (Fig.76). The voltage readings and the differences are so small that special circuitry is necessary to read them, but the readings can be used as a measure of crack height.

With an AC instrument the current is carried in only a thin layer at the metal surface, due to the skin effect. Thus, only a few amperes are necessary to give a measurable potential difference. At a frequency of 5MHz the skin depth is 0.13mm, and a current of 1A will produce a measurable surface voltage (e.g. $1\mu V$). It can be shown[4] that, by making two measurements of the field and knowing the probe spacing, it is possible to determine the crack depth without prior calibration. There are also DC instruments available which use the potential-drop technique. With DC there is considerable penetration of the

76 Four-probe contact potential-drop method. The outer probes produce a current flow; the inner pair measure the voltage.

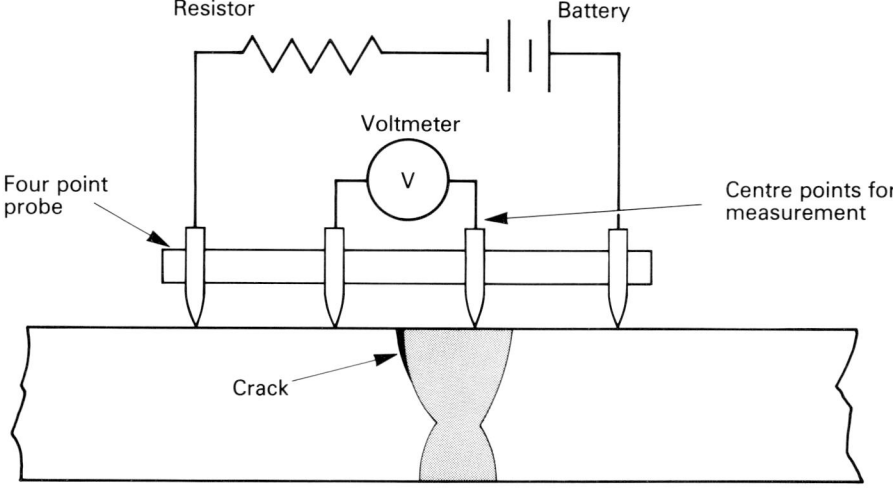

current into the specimen, so that larger currents must be used and contact problems can occur. One type of DC potential-drop equipment uses a four-point contact probe.

Metal sorting

A variant of the eddy current technique can be used to test or compare ferromagnetic components for properties such as composition, hardness, internal stresses and conditions such as case-hardening depth.[5]

Two identical coil assemblies, either the encircling or probe types, are set up at right angles to one another (so that the field of one does not affect the other) and both are fed with a 50Hz AC voltage so that the two currents are 180° out of phase. The coils are connected to the Y-plates of an oscilloscope, with the X-plates being controlled by a time base, and the two signals are superimposed. The two phases then cancel out and a horizontal straight line is displayed.

If, now, a test sample is introduced to one of the coils, the material undergoes magnetic hysteresis, the loop of which is modified by the action of induced eddy currents, and the oscilloscope trace assumes a shape which depends upon the electrical conductivity, magnetic permeability and dimensions of the test sample. If a specimen is put into each coil, the trace is horizontal for two identical specimens but assumes a characteristic shape for any variation between the two samples.

Various forms of coil may be used to test different shapes of specimen, such as bar stock and tubing, and to sort out different alloys. It is important to note that, if material is to be sorted according to alloy composition, the physical dimensions of the specimens must be accurately controlled.

References

1   BS 3889: Part 2A: 1986: Automatic eddy current testing of wrought steel tubes. Publ British Standards Institution, 1986.
2   Ibid: Part 2B: 1982: Eddy current testing of non-ferrous tubes. Publ British Standards Institution, 1982.
3   BS 3683: Part 5: 1983: Eddy current flaw detection. Publ British Standards Institution, 1983.
4   Dover W D and Collins R: *Brit J NDT* 1980 **22** (6) 291.
5   Blitz J, King W G and Rogers D G: *Electrical, Magnetic and Visual Methods of Testing Materials*. Publ Butterworths, London, 1969.

7 Other methods of NDT

Many other NDT methods are used on a limited scale for specific applications: some of these methods have potential application to weld inspection under particular conditions.

Acoustic emission

This is a quite different concept from the foregoing methods of NDT already detailed in that no external signal is put into the specimen to be tested. It is an entirely passive test, but it can be applied only to a specimen which is subjected to stress. However, this stress need not be external: it can be internal.

At certain stress levels a solid is subject to plastic deformation and fracture, and energy is released. Some of this is converted to elastic waves that are emitted as discrete pulses of acoustic energy, which travel through the specimen and can be detected by sensitive transducers placed on the surface of the specimen. This phenomenon of sound generation in materials under stress is called 'acoustic emission', or AE. As materials approach fracture, the quantity of AE generally increases very rapidly in both intensity and number of pulses. The sources of AE are very varied and not yet fully understood, and it is important to realise that only a proportion of the energy released on fracture or crack propagation is emitted as AE. The special point about AE methods, compared with most other forms of NDT, is that the receiving transducer need not be particularly near the source of the emissions, so that the test is not localised. Also, the AE is detected in real time, i.e. as it occurs. The emissions will be attenuated in intensity and dispersed between the source of the AE and the detector.

The amount of emission produced by different engineering materials varies enormously: some metals are said to be 'quiet', which has led to failure to detect incipient fracture in some applications and caused controversy over the efficacy of AE techniques. Moreover, background noise produced during in-service testing may mask AE, but nevertheless it remains a potentially powerful method of monitoring large stressed structures (including welded fabrications) during their service life or, for example, during hydrostatic proof testing.

Another important physical phenomenon which must be considered in AE work is known as the Kaiser effect. If a polycrystalline material such as a metal is stressed and then relaxed, no new AE occurs when the specimen is re-stressed until the previous maximum stress has been exceeded.

77 Block diagram of
acoustic emission (AE)
equipment.

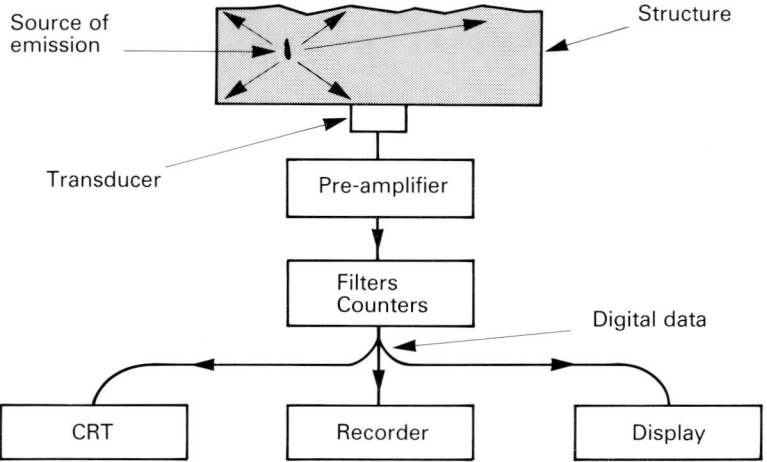

Equipment

The basic requirements are very simple (Fig.77): a very sensitive piezoelectric transducer, amplifiers and counters. Costs, however, mount rapidly as recording and data analysis systems are added, and most applications use several transducers, each with its own circuits. The transducer is bonded to the surface of the specimen and connects to the electronic equipment by a special 'super-screened' cable. Normally, wide band transducers operating over the range 200-1000kHz are used, but transducers with peak sensitivity at particular frequencies are also used as these can be intrinsically more sensitive.

Information on AE can be recorded as rate of emissions received, frequencies within the pulse, amplitudes or energies. The simplest method of characterising a series of pulses is known as 'ring-down counting'. This method counts the number of times per second that the AE amplitude exceeds a pre-set voltage, but does not take account of the point that a large amplitude signal will usually last longer than one of low amplitude, so that the count obtained is biased towards large amplitude pulses. Energy analysis of the pulses, by measuring the area under the pulse envelope on the amplitude/time curve, or by squaring the initial pulse amplitude, would seem to be a more logical method of recording AE data, and most equipment nowadays is capable of this technique.

The limitation of energy analysis methods is that the pulses are attenuated during their travel from the AE source to the transducer, this attenuation being frequency-dependent, so that the form of the pulse, as recorded, is not the same as the emitted pulse.

Frequency filtering and digital methods of recording are commonly used, and the latter allows much larger amounts of data to be collected and analysed.

Applications of acoustic emission

There are three major groups of applications of AE: the location of defects; the detection of crack propagation; and the monitoring of processes in real time. The first uses several transducers, records the time of arrival of a signal at each one, and by triangulation determines the location of a source of emission of AE, which can be calculated with the help of computer programs. The region of the specimen from which AE has originated can then be examined for defects by other NDT methods such as ultrasonic testing.

The second group of applications depends on the recording of AE as a function of time and is designed to detect a gradual or sudden increase in emission, which would indicate the propagation of a defect, such as a crack or the development of new flaws in the structure.

In the third group of applications, the main difficulty is the severe interference from noise generated by the process itself, such as metal transfer or slag cracking, as well as from electrical interference. Considerable success, however, has been achieved in spot welding monitoring and in monitoring delayed cracking in welds. Even with the more conventional welding processes, some workers recommend AE as a valuable aid to ensuring good welds.[1]

For in-service surveillance of welded structures, the results have been varied. Some workers consider that 'quiet steels' nullify the technique; others consider that better amplification circuitry and background noise suppression will still allow some AE to be detected, and considerable success has been achieved in trials where failure eventually arose from stress corrosion cracking.

Optical holography

Optical holography is a recording method which uses intensity, wavelength and phase of light reflected from an object to display minute deformations or variations in deformation across a surface under stress. This is used to show the presence of internal defects or structural variations under the surface, but has not so far found much application to weldments. The light used to illuminate the surface of the specimen must be coherent, which means that it must also be monochromatic, and the only practical source is a laser. Each type of laser emits a characteristic wavelength, e.g. a helium-neon laser emits 632.8nm; a ruby laser emits 694.3nm.

Holographic interferometry is the method used to measure minute deformations and consists in 'holographing' the object for half the normal exposure time, then altering the stress on the object and making a second half-exposure. Each holographic exposure photographs the object illuminated in laser light with a split beam, so that half the light reaching the photographic plate travels direct to the plate (the reference beam) and the other half is reflected off the object (Fig.78). These two beams interfere, since both originate from the same coherent source, resulting in a hologram, which usually has a speckled appearance, apparently unrelated to the usual appearance of the specimen. This photographic hologram must be produced on a special photographic material of very high resolving power and with a colour sensitivity to match the laser light used.

After processing, the silver image is bleached out of the holographic plate leaving a 'phase hologram', and the image is reconstructed from this phase-only hologram by illumination in laser light. The reconstructed image of the double-exposure shows bands of interference fringes, the number and spacing of which indicate the amount of object distortion stemming from the change in stress. Thus, local variations in the distribution of these interference fringes indicate the presence of internal flaws or structural changes. The interference effects are entirely independent of the surface roughness.

If the hologram reconstruction process is carried out with the specimen still in its original position, the holographic image superimposes exactly on to the

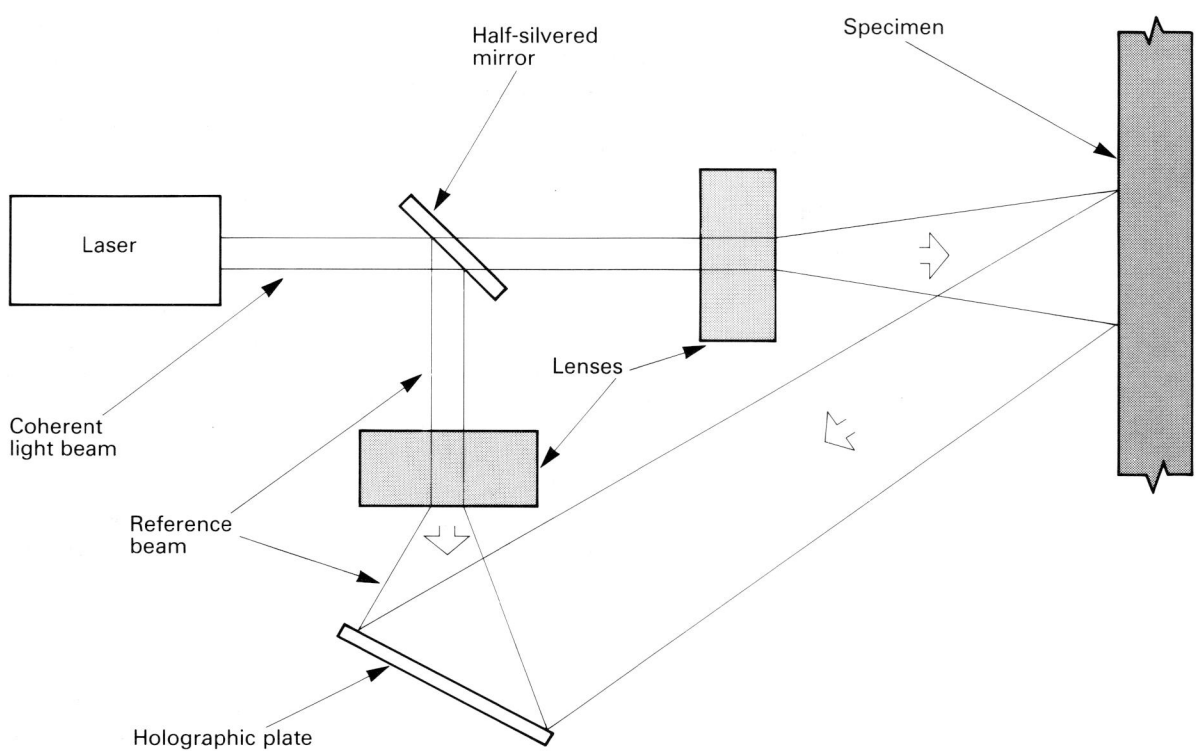

78 Optical holography for surface deformations.

specimen and any fringes due to distortion can be seen 'live'. To date, the most important industrial application of holography has been in the inspection for poor adhesion of the bonding of laminated panels for aircraft, such as those with honeycomb structures, but there are clearly potential applications in the examination of fabricated structures for distortion following the onset, for example, of fatigue or creep.

Speckle interferometry

When a surface is illuminated by laser light it has a 'speckled' appearance, and if this speckle pattern is photographed it can also be used for NDT purposes. If the processed photographic image of the speckle pattern is replaced in its original position, it can act as a negative mask, so that if the surface of the specimen is misaligned in any way, as a result of local distortion, the negative and the speckle pattern will not match perfectly, and the nature of the resultant speckle can be used as a measure of specimen distortion. This method is inherently simpler than holographic interferometry.

Ultrasound imaging systems

In Chapter 4, ultrasonic pulse echo methods of flaw detection were described in detail as being among the most important methods for the NDT of welds. There is, however, another much less widely employed use of ultrasound in which the transmitted beam from a wide field source is detected over an extended area and used to form an ultrasound image, analogous to an X-ray image. The attraction of the method is that ultrasound can be transmitted through large thicknesses of fine grain metal and that no ionising radiation hazard is involved. Several systems have been devised,[2] including ultrasonic image converter tubes, computer reconstruction methods and acoustic holography.

It must be remembered, however, that the wavelength of the ultrasonic waves used for metal inspection is of the order of 1-4mm, so that high resolution images cannot be expected.

Ultrasound imaging tubes

Ultrasound imaging tubes have been built which use an extended-area piezoelectric pick-up surface which is scanned by an electron beam, exactly as in a TV camera tube. As such tubes have to be used in a liquid bath for coupling purposes, an acoustic reducing lens can be used so that the pick-up surface need not be the full size of the specimen. Alternatively, a specimen scanning system can be used (Fig.79). The resolution is largely limited by the ultrasonic wavelength in the piezoelectric material of the pick-up surface (for example, 1.4mm at 2MHz). If the pick-up surface is made thinner, to utilise higher ultrasonic frequencies, mechanical strength problems arise. This type of ultrasound converter tube has been applied to spot weld inspection.[3]

Multiple transducer imaging systems

If the ultrasonic transducer is built up of a series of elements and each element is fed separately with generating pulses, by phasing these pulses appropriately the ultrasonic beam shape and the beam direction can be varied and controlled. With similar electronic control of the receiver pulses, an ultrasonic image can be built up. This technique has developed rapidly in recent years for medical ultrasonic imaging, but has not yet found much industrial application.

A simpler arrangement, employing the same principle, is to move a probe over the surface and build up an image on a storage display tube. Many systems of this general type are under development but it is not yet possible to predict which

79 Ultrasonic imaging using a converter tube in an immersion tank.

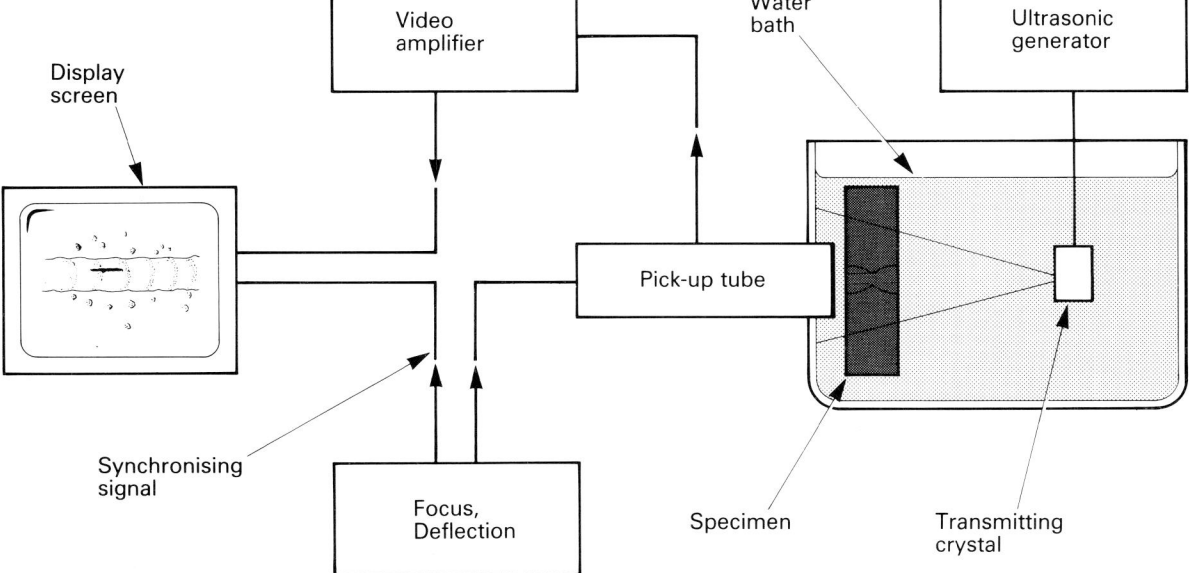

designs will become important for weld inspection. However, any system which enables a defect to be identified by visualising its surface contour is of obvious potential importance. This technique seems likely to be replaced by computer image reconstruction from data converted to digital form.

Acoustic holography

Holograms can be made with ultrasonic waves in the same manner as with light, by using interference between a reference beam and the object beam, provided these are coherent. The resulting image can be displayed or digitally stored for computer processing.

Because of coupling problems at surfaces, it is essential that the specimen and the ultrasonic beams are used in an immersion tank. Figure 80 shows one possible arrangement, in which the standing waves on the water surface form the ultrasonic hologram, and by illuminating this with a beam of coherent (laser) light, suitably focused, the image is made visible.

Alternatively, the ultrasonic hologram can be produced by using a single ultrasonic beam to cover the specimen, a small scanning transducer to detect this and an electronic reference beam. Variants on this system using a pulsed ultrasonic beam, with the same transducer acting as transmitter and receiver, are also possible.[4]

80 Ultrasonic holography, using an immersion tank with water surface levitation and optical visualisation.

Some workers have attempted to use water column probe coupling to eliminate the need for an immersion tank, and this technique coupled with an ultrasonic reference beam and computer data processing (numerical reconstruction) would seem most likely to lead to a practical industrial inspection

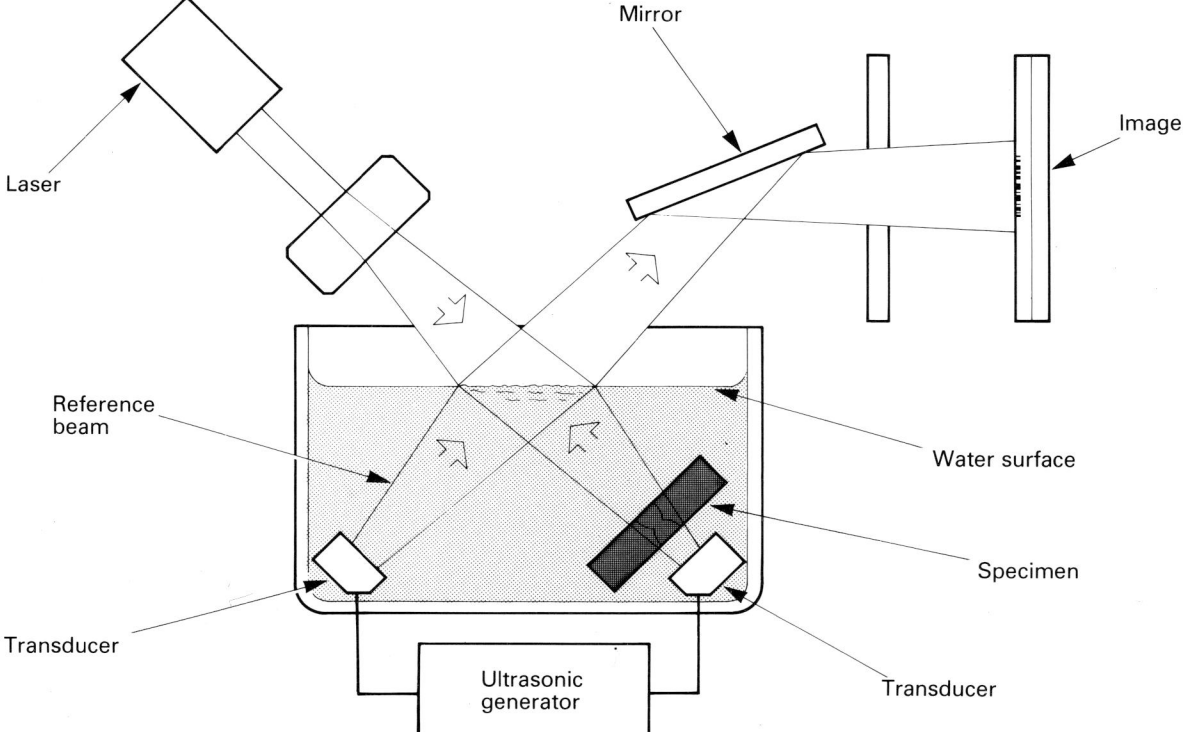

system. The attraction of ultrasonic holography is that, if an image of the flaw can be seen, it is more likely to be identifiable. Secondly, there is the possibility of greater resolution than with conventional ultrasonic testing using the same frequency. Much more development in the technique is likely during the next ten years.

References

1   Williams R V: 'Acoustic emission.' Publ Adam Hilger Ltd, Bristol, 1980, Ch 5.
2   Jacobs J E: 'Research techniques in non-destructive testing.' Vol 1. Publ Academic Press, London, 1970, Ch 3.
3   Matting N and Wilkens G: *Ultrasonics* 1965 **3** (2) 161.
4   Aldridge E E: 'Research techniques in nondestructive testing.' Vol 1. Op cit, Ch 5.

In conclusion

This book has described both the established methods of non-destructive testing applicable to welded joints and those under development, and pointed out the advantages and disadvantages of each.

Radiography, although slow and expensive, particularly on thick metals, produces an image directly related to the flaws, which is usually easy to interpret; but it may not detect very fine cracks.

Ultrasonic testing by established manual methods is reliant on the skill of the operator for correct calibration, application and interpretation of the displayed signals to estimate the nature and size of the flaws. Potentially, *manual testing* can detect the finest cracks, although the realisation of this in practice is totally dependent on the operator. Ultrasonic testing may be unsuitable for coarse-grained materials such as welds in some austenitic steels. *Automated equipment* with several probes, computer control and computer read-out is being rapidly developed to produce consistent results on repetition work, but at a high cost.

Magnetic particle crack detection, applicable only to ferromagnetic materials and *penetrant testing,* are both very sensitive, cheap methods of detecting surface breaking cracks. Penetrant testing requires very careful cleaning of the weld surface prior to inspection. One or other of these two methods should be used to supplement radiographic and ultrasonic testing. The Table below summarises these advantages and disadvantages:

Method	Advantages	Disadvantages
Radiography	Direct image of flaws Applicable to all metals	Slow, high cost Safety hazard Limited ability to detect fine cracks
Ultrasonic testing (manual)	Can detect fine cracks	No direct flaw image Totally reliant on operator Not suitable for coarse-grained metals
Ultrasonic testing (automatic)	Can detect fine cracks Can produce direct image Consistent results	High cost Best suited for repetition work Not suitable for coarse-grained metals
Magnetic-particle crack detection	Sensitive to fine cracks Inexpensive	Ferromagnetic materials only Surface breaking defects only
Penetrant testing	Any material Inexpensive Sensitive to fine cracks	Surface breaking defects only Needs prior surface cleaning

Currently, non-destructive testing can find most defects in welded joints, ensuring safe long term operation of plant and structures; but this is achieved at considerable expense — and current developments are directed at producing consistent reliable testing at reasonable cost.

Although some of the newer methods of NDT will find immediate limited application to weld inspection, particularly on special types of welding such as friction welds and welds between dissimilar metals, it seems likely that in the foreseeable future non-destructive testing of welds will continue to depend on ultrasonic methods, radiography and one or other of the two surface crack detection methods — magnetic particle or penetrants.

Index